Advanced Energy
Design Guide
for
Small Retail Buildings

This publication was developed under the auspices of ASHRAE Special Project 110.

ADVANCED ENERGY DESIGN GUIDE—SMALL RETAIL COMMITTEE

Merle McBride, *Chair*

Ron Jarnagin
Vice Chair

Ronald Kurtz
IESNA Representative

Don Colliver
Steering Committee Ex Officio

Michael Lane
IESNA Representative

Donald Brundage
ASHRAE TC 2.8 Representative

Harry Misuriello
ASHRAE TC 7.6 Representative

Charles Culp
ASHRAE TC 9.5 Representative

Dan Nall
USGBC Representative

Jay Enck
USGBC Representative

Paul Torcellini
Consultant

Katherine Hammack
ASHRAE SSPC 90.1 Representative

Bruce Hunn
ASHRAE Staff Liaison

David Hartke
AIA Representative

AEDG STEERING COMMITTEE

Don Colliver, *Chair*

Jeffrey Levine
AIA

Ron Majette
Ex Officio

Terry Townsend
ASHRAE

John Hogan
Consultant (ASHRAE TC 2.8)

Rita Harrold
IESNA

Harry Misuriello
Consultant (ASHRAE TC 7.6)

Brenden Owens
USGBC

Jerry White
Consultant (ASHRAE Standard 90.1)

Advanced Energy Design Guide

for

Small Retail Buildings

Achieving 30% Energy Savings Toward a Net Zero Energy Building

American Society of Heating, Refrigerating and Air-Conditioning Engineers
The American Institute of Architects
Illuminating Engineering Society of North America
U.S. Green Building Council
U.S. Department of Energy

ISBN 1-933742-06-2

Cover design by Emily Luce, Designer.
Cover photograph courtesy of Communities by Design, American Institute of Architects.
Illustrations by Richard J. Vitullo, AIA, Vitullo Architecture Studio, P.C.

Library of Congress Cataloging-in-Publication Data

Advanced energy design guide for small retail buildings : achieving 30% energy savings over ANSISSHRAE/IESNA standard 90.1-1999. / American Society of Heating, Refrigerating and Air-Conditioning Engineers ... [et al.].
   p. cm.
 Summary: "Second in series that provides recommendations for achieving 30% energy savings over minimum code requirements of ANSI/ASHRAE/IESNA Standard 90.1-1999. Focuses on small retail buildings up to 20,000 sq. ft. Recommendations for the 8 US Climate Zones allow owners/designers to achieve advanced levels of savings without resorting to calculations or analysis"--Provided by publisher.
 ISBN 1-933742-06-2 (softcover)
 1. Buildings--Energy conservation. I. American Society of Heating, Refrigerating and Air-Conditioning Engineers.

TJ163.5.B84A27 2006
725'.210472--dc22
        2006027504

ASHRAE Staff

SPECIAL PUBLICATIONS

Mildred Geshwiler
Editor

Christina Helms
Associate Editor

Cindy Sheffield Michaels
Assistant Editor

Michshell Phillips
Administrative Assistant

PUBLISHING SERVICES

David Soltis
Manager

Jayne Jackson
Publication Traffic Administrator

PUBLISHER

W. Stephen Comstock

Contents

Acknowledgments

The *Advanced Energy Design Guide for Small Retail Buildings* is the second in a series of anticipated Guides that will address many types of buildings. A huge debt of gratitude is extended to the authors of the first Guide on small office buildings because they paved the way and defined the basic structure, content, and format of the Guides as well as the procedures for the reporting and the reviews. Following in their footsteps has provided consistency among these two Guides in addition to being a tremendous time saver.

Continuity with the first Guide on small office buildings was further maintained because many of the same organizational partners were involved, as were several of the project committee members. ASHRAE was again the lead organization, with full support from the American Institute of Architects (AIA), the Illuminating Engineering Society of North America (IESNA), the U.S. Green Building Council (USGBC), and the U.S. Department of Energy (DOE). Individuals from each of these organizations were members of the Steering Committee. Under the leadership of 2002–2003 ASHRAE President Don Colliver, their contributions were significant in terms of the direction and oversight they provided to the ASHRAE Special Project 110 Committee (SP-110). Members on the project committee came from the partner organizations and ASHRAE Standing Standards Project Committee 90.1 (SSPC 90.1) and the ASHRAE Technical Committees on Building Environmental Impact and Sustainability (TC 2.8) and System Energy Utilization (TC 7.6). These members served not only on the project committee but also as liaisons to their respective organizations, and they coordinated the multiple technical reviews.

The best way to characterize the project committee members was their dedication and professionalism. Mr. Ron Jarnagin of the Pacific Northwest National Laboratory served as vice-chair. He was a valued asset since he had been the chair for the first Guide on small office buildings. His leadership experience was utilized to avoid previous challenges and provide smooth sailing. All of the project committee members sacrificed five full weekends for meetings, conducted three external reviews, responded to all of the remarks, participated in nine conference calls, identified many case studies, participated in monitoring focus groups, developed all of the recommendations, and prepared the accompanying text. This was accomplished within a time period of eight months, which has become the default length of an ASHRAE year for the development of these Guides. All of the project committee members were highly motivated and skilled, which made my job as chair much easier. All I had to do was define the end goal and try to stay out of their way.

In addition to the voting members on the committee, there were many other individuals who contributed to the success of this Guide. The specific individuals and their contributions were: Paul Torcellini of the National Renewable Energy Laboratory for his technical expertise and multiple case study examples; Bing Liu of Pacific Northwest National Laboratory for the simulation runs and results; Bruce Hunn of ASHRAE Staff for his technical support, recording of the meeting notes, and serving as gracious host of all the meetings at ASHRAE Headquarters; Micki Geshwiler of ASHRAE Staff for taking the lead on the focus groups; Cindy Sheffield Michaels of ASHRAE Staff for editing and layout of the book; and John Hogan for his expertise on building envelopes. This Guide could not have been developed without their contributions.

As chair of the committee, I am very proud of the Guide that the project committee has developed, and each member can take pride in their individual contributions.

Merle McBride
SP-110 Chair
August 2006

Abbreviations and Acronyms

A	=	area, ft^2
ACCA	=	Air Conditioning Contractors of America
AEDG-SR	=	*Advanced Energy Design Guide for Small Retail Buildings*
AFUE	=	annual fuel utilization efficiency, dimensionless
AIA	=	American Institute of Architects
ASHRAE	=	American Society of Heating, Refrigerating and Air-Conditioning Engineers
ASTM	=	American Society for Testing and Materials
ANSI	=	American National Standards Institute
Btu	=	British thermal unit
C	=	thermal conductance, $Btu/h \cdot ft^2 \cdot °F$
c.i.	=	continuous insulation
Cx	=	commissioning
CxA	=	commissioning authority
cfm	=	cubic feet per minute
CMH	=	ceramic metal halide
COP	=	coefficient of performance, dimensionless
CRI	=	Color Rendering Index
CRRC	=	Cool Roof Rating Council
D	=	diameter, ft
DL	=	*Advanced Energy Design Guide* code for "daylighting"
DOE	=	U.S. Department of Energy
E_c	=	efficiency (combustion), dimensionless
EF	=	efficiency
EIA	=	Energy Information Administration
E_t	=	efficiency (thermal), dimensionless
EER	=	energy efficiency ratio, $Btu/W \cdot h$
EF	=	energy factor

EL	=	*Advanced Energy Design Guide* code for "electric lighting"
EN	=	*Advanced Energy Design Guide* code for "envelope"
EX	=	*Advanced Energy Design Guide* code for "exterior lighting"
F	=	slab edge heat loss coefficient per foot of perimeter, Btu/h·ft·°F
FCU	=	fan-coil unit
GC	=	general contractor
Guide	=	*Advanced Energy Design Guide for Small Retail Buildings*
HC	=	heat capacity, Btu/ft^2·°F
HSPF	=	heating season performance factor, Btu/W·h
HV	=	*Advanced Energy Design Guide* code for "HVAC systems and equipment"
HVAC	=	heating, ventilating, and air-conditioning
IESNA	=	Illuminating Engineering Society of North America
in.	=	inch
IPLV	=	integrated part-load value, dimensionless
kBtuh	=	thousands of British thermal units per hour
kW	=	kilowatt
LBNL	=	Lawrence Berkeley National Laboratory
LED	=	light-emitting diode
LPD	=	lighting power density, W/ft^2
N/A	=	not applicable
NBI	=	New Buildings Institute
NEMA	=	National Electrical Manufacturers Association
NFRC	=	National Fenestration Rating Council
NREL	=	National Renewable Energy Laboratory
NZEB	=	net zero energy buildings
O&M	=	operation and maintenance
OPR	=	Owner's Project Requirement
PL	=	*Advanced Energy Design Guide* code for "plug loads"
ppm	=	parts per million
PV	=	photovoltaic
QA	=	quality assurance
R	=	thermal resistance, h·ft^2·°F/Btu
RPI	=	Rensselaer Polytechnic Institute
SEER	=	seasonal energy efficiency ratio, Btu/W·h
SHGC	=	solar heat gain coefficient, dimensionless
SRI	=	Solar Reflectance Index, dimensionless
SSPC	=	standing standards project committee
SWH	=	service water heating
TAB	=	test and balance
TC	=	technical committee
U	=	thermal transmittance, Btu/h·ft^2·°F
UPS	=	uninterruptible power supply
USGBC	=	U.S. Green Building Council
VLT	=	visible light transmittance
W	=	watts
w.c.	=	water column
WH	=	*Advanced Energy Design Guide* code for "water heating systems and equipment"

Introduction

1

The *Advanced Energy Design Guide for Small Retail Buildings* (AEDG-SR; the Guide) is intended to provide a simple approach for contractors and designers who create retail buildings up to 20,000 ft^2. Application of the recommendations in the Guide should result in small retail buildings with 30% energy savings when compared to those same retail buildings designed to the minimum requirements of ANSI/ASHRAE/IESNA Standard 90.1-1999. This document contains recommendations and is *not* a minimum code or standard. It is intended to be used in addition to existing codes and standards and is not intended to circumvent them. This Guide represents *a way*, but *not the only way*, to build energy-efficient small retail buildings that use significantly less energy than those built to minimum code requirements. The recommendations in this Guide provide benefits for the owner while maintaining quality and functionality of the space.

This Guide has been developed by a committee representing a diverse group of energy professionals drawn from the American Society of Heating, Refrigerating and Air-Conditioning Engineers (ASHRAE), the American Institute of Architects (AIA), the U.S. Green Building Council (USGBC), and the Illuminating Engineering Society of North America (IESNA). To quantify the expected energy savings, these professionals selected potential envelope, lighting, HVAC, and service water heating (SWH) energy-saving measures for analysis. These included products that were deemed to be both practical and commercially available. Although some of the products may be considered premium, products of similar performance are available from multiple manufacturers. Each set of measures was simulated using an hourly energy analysis program for two retail prototype buildings in representative cities in various climates. Simulations were run for reference buildings (buildings designed to Standard 90.1-1999 criteria) compared to buildings built using recommendations contained in this Guide to determine whether the expected 30% savings target was achieved.

The scope of this Guide covers small retail buildings up to 20,000 ft^2 in size that use unitary heating and air-conditioning equipment. Buildings of this size with these HVAC system configurations represent a significant amount of commercial retail space in the United States. This Guide provides straightforward recommendations and how-to tips to facilitate its use by anyone in the construction process who wants to produce more energy-efficient buildings.

In general, this Guide addresses typical retail building uses: retail (other than shopping malls); strip shopping centers; automobile dealers; building material, garden supply, and hardware stores; department stores; drugstores; furniture, home equipment, and home furnishing stores; liquor stores; and wholesale goods (except food). The Guide *excludes* certain building uses such as car washes; laundry and dry-cleaning establishments; gasoline service stations; motor vehicle repair, service, and maintenance buildings; personal service establishments (barbers, hair dressers, manicurists, etc.); and other facilities that have significant point source heat or pollutant generation. The Guide also excludes treatment of food service facilities such as delicatessens and restaurants. Also excluded from the Guide are "built-up" HVAC systems using chillers and chilled-water systems.

As an added value for designers and contractors, this Guide features technology examples and case studies of energy-efficient buildings. The case studies demonstrate that effectively addressing environmental challenges can also result in the creation of good, often excellent, architecture. The examples illustrate how energy considerations have been incorporated in various design strategies and techniques. However, the example buildings may incorporate additional features that go beyond the scope of the recommendations of the Guide.

It is hoped that the Guide will result in a more sustainable environment for society. The energy savings target of 30% is the first step in the process toward achieving a net zero energy building (NZEB), which is defined as a building that, on an annual basis, draws from outside sources equal or less energy than it provides using on-site, renewable energy sources. ANSI/ASHRAE/IESNA Standard 90.1-1999, the energy conservation standard published at the turn of the millennium, provides the fixed reference point for all the Guides in this series. The primary reason for this choice as the reference point is to maintain a consistent baseline and scale for all the 30% AEDG series documents. A shifting baseline between multiple documents in the AEDG series would lead to confusion among users about the level of energy savings achieved. The average energy savings for this Guide over all climates and buildings analyzed was approximately 37% when compared to Standard 90.1-1999. However, it is interesting to see what the energy savings of the Guide would be relative to Standard 90.1-2004, which has reduced the lighting power densities and improved efficiency levels for the cooling equipment. Using Standard 90.1-2004 as the basis, the average energy savings for this Guide over all climates and buildings analyzed was approximately 30%, or seven percentage points less than when compared to Standard 90.1-1999.

Plans are in place for development of additional AEDG documents in this decade that will assist users in achieving 50% and 70% energy-saving levels as milestones toward the NZEB goal. The ultimate goal of the participating organizations is to assist in the design and construction of NZEBs.

CONTENTS

Chapter 2 of this Guide contains a chart that walks the user through the design process of applying the recommendations in this Guide, while Chapter 3 provides the actual recommendations for a way to meet the 30% energy savings goal. Chapter 3 includes eight recommendation tables, which are broken down by building component and organized by climate according to the eight climate zones (and specific counties within each climate zone) identified by the U.S. Department of Energy (DOE). The user should note that the recommendation tables do not include all of the components listed in Standard 90.1-1999 since the Guide focuses only on the primary energy systems within a building. Chapter 4 provides technology examples and case studies of actual energy-efficient buildings and systems. Chapter 5 provides essential guidance in the form of concise how-to tips to help the user understand and apply the recommendations from this Guide. Additional "bonus savings" strategies are also found in Chapter 5.

This Guide includes specific recommendations for energy-efficient improvements in the following technical areas to meet the 30% energy savings goal:

- Building Envelope
  - Roofs
  - Walls

- Floors
- Slabs
- Doors
- Vertical Glazing
- Skylights
- Lighting
  - Daylighting
  - Interior Electric Lighting
  - Lighting Controls
- HVAC Equipment and Systems
  - Cooling Equipment Efficiencies
  - Heating Equipment Efficiencies
  - Supply Fans
  - Ventilation Control
  - Ducts
- Service Water Heating
  - Equipment Efficiencies
  - Pipe Insulation

In addition, "bonus savings" strategies to improve energy efficiency beyond the 30% energy savings level are included for:

- Exterior Façade Lighting
- Parking Lot Lighting
- Plug Loads and In-Store Illuminated Displays
- Exterior "Internal" Illuminated Signage, Façade Applied or within Storefront Windows

Quality assurance (QA) and commissioning (Cx) are also covered in Chapter 5.

HOW TO USE THIS GUIDE

There are numerous ways to use this Guide effectively consistent with the background and knowledge of the user—some may turn immediately to the climate-specific recommendations; others may choose to first understand how energy goals fit into the design process. In addition, this Guide provides recommendations that would assist the user in achieving energy efficiency credits for LEED or other building energy rating systems. The authors of this Guide suggest the following approach:

- Review Chapter 2 to understand how energy efficiency goals relate to the stages of building design. The flow charts, tables, and checklists in Chapter 2 can be used to lead, communicate, and manage the design and construction of energy-efficient buildings.
- Review Chapter 3 for specific recommendations to achieve the 30% energy savings level in your climate zone. These criteria provide *a way* to achieve the 30% savings goal and also serve as a starting point to further refine the energy design. The authors realize that designers typically don't receive sufficient design fees to perform energy design optimization. Therefore, the contents of this chapter can serve as a staring point to meet specific requirements of a particular project.
- Review the Chapter 4 case studies to assure you and your team that other designers and builders have successfully used these and similar techniques to build energy-efficient buildings in the real world. In fact, a number of the case study buildings have won awards or achieved peer recognition for their energy efficiency attributes.
- Use Chapter 5 to detail how the Chapter 3 recommendations are applied. Use the how-to tips, cross-referenced to the recommendation tables, to apply best practices (as well as cautions to avoid known problems in energy-efficient construction) to the specific circumstances of your project. Also, consider the recommendations for "bonus savings" from energy-efficient appliances and cord-connected equipment as well as exterior lighting controls.

Integrated Process for Achieving Energy Savings 2

This chapter of the AEDG-SR provides resources for those who want to understand and adopt an overall *process* for designing, constructing, and operating energy-efficient small retail buildings. The resources listed below are above and beyond the straightforward presentation of recommendations in Chapter 3 and the how-to tips in Chapter 5 that lead to energy savings of 30% beyond Standard 90.1-1999. The resources are:

- *A narrative discussion of the design and construction process that points out the opportunities for energy savings in each phase.* It further explains the steps that each team member or discipline should take to identify and implement energy savings concepts and strategies. It also includes a discussion on how the quality assurance (QA) measures are worked into the process at each phase and how some of these measures can be used by the owner to maintain energy performance for the life of the facility.
- *A reference table or matrix that leads the Guide's user through the process of identifying and selecting energy-saving measures to meet major energy design goals.* This information is presented in Table 2-5, which ties together detailed strategies, recommendations for meeting the 30% energy use reduction target, and related how-to information.

The following presentation of an integrated process for achieving energy savings in small retail buildings is valuable for designers and builders who want to augment and improve their practices so that energy efficiency is deliberately considered at each stage of the development process from project conception through building operation. Commissioning (Cx) begins in the early stage of design and continues through operation and is an integral part of each phase. These stages are shown in Figure 2-1.

The key benefits of following this integrated process include:

- Understanding the specific step-by-step activities that owners, designers, and construction team members need to follow in each phase of the project's delivery, including communication of management, design, construction, QA, Cx, operation and maintenance (O&M), and occupancy functional requirements an owner should follow to maintain the specified energy performance of the facility.
- Identifying energy efficiency goals and selecting design strategies to achieve those goals.

Figure 2-1. Stages of design.

- Incorporating QA, including building Cx procedures, into the building design and delivery process to ensure that energy savings of recommended strategies are actually achieved and that specific documentation needed to maintain energy performance is provided to the owner.
- Owner understanding of the ongoing activities needed to help ensure continued energy performance for the life of the facility, resulting in lower total cost of ownership.

Users of this Guide who follow the recommended process in their design and construction practice will benefit from achieving the goals of enhanced energy savings.

1. DESIGN PHASE (INCLUDING PLANNING AND PRE-DESIGN)

Documentation of the adopted energy goals and general strategies begin in the Design Phase. In small retail, this typically involves writing the Owner's Project Requirement (OPR) document, a brief, two-page description that includes the project's energy goals. The OPR document guides the team and provides a guide to be used during the design and construction of the project.

This Guide emphasizes goals that relate to the energy uses that can produce the largest savings. Interior lighting plays a major role in retail building energy use. Both Standard 90.1-1999 and the Guide recognize differences between general lighting and accent lighting. General lighting applies to the entire building. Accent lighting is additional interior lighting that is allowed provided it is specifically designed and directed to merchandise and is automatically controlled separately from the general lighting. Standard 90.1-1999 specifies lighting power densities (LPDs) for general lighting and two levels of accent lighting. The Guide specifies lower LPD for general lighting and has four levels of accent lighting. Comparisons of the LPDs between Standard 90.1-1999 and this Guide are presented in Figure 2-2 for five typical retail buildings with various merchandise and 70% display floor area.

In addition to lighting, differences in building application, climate, and even orientation will impact the selection of various energy goals and strategies. As an example, Figures 2-3 and 2-4 show energy use mixes for three stores in a 7,500 ft^2 strip mall in two locations, Phoenix and Chicago. Store 1 is 25 ft wide by 75 ft deep (1875 ft^2) for general merchandise with 100% general lighting (1.82 W/ft^2 baseline and 1.28 W/ft^2 advanced), store 2 is 50 ft wide by 75 ft deep (3750 ft^2) with 75% general lighting and 25% accent lighting (2.94 W/ft^2 baseline and 1.69 W/ft^2 advanced), and store 3 is 25 ft wide by 75 ft deep (1875 ft^2) with 50% general lighting and 50% accent lighting (4.55 W/ft^2 baseline and 2.33 W/ft^2 advanced). These figures show that cooling and lighting energy predominates in Phoenix; thus, in that climate zone, the goals and strategies relating to cooling and lighting should receive the highest priority. Conversely, in Chicago, the goals and strategies relating to heating and lighting should receive the highest priority. Also, in the "Bonus Savings" section of Chapter 5, specific examples provide methods to achieve savings above the Chapter 3 requirements.

In the Design Phase, the team integrates the energy goals into building plans, sections, details, and specifications. The sequence of many design decisions, such as building and glazing orientation as well as other strategies identified in this chapter, has a major impact on energy efficiency. These decisions must, therefore, be made much sooner in the process than is typically done. The steps listed in Table 2-1, presented in sequence, identify the appropriate time in the process to apply specific recommendations from this Guide.

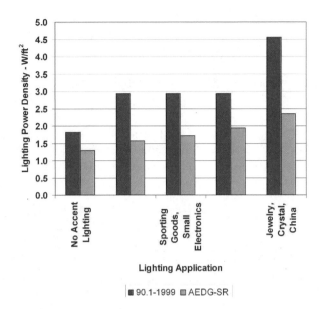

Figure 2-2. Lighting power densities.

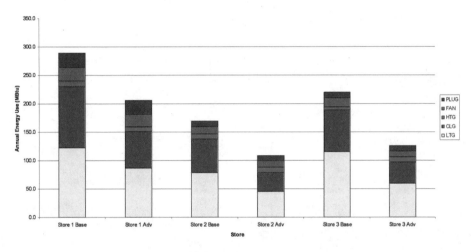

Figure 2-3. Strip mall energy use for three stores in Phoenix.

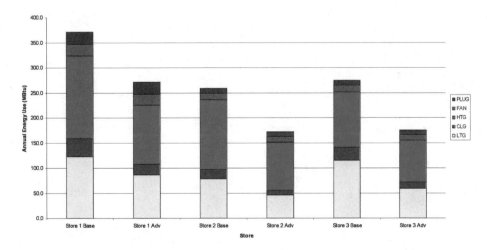

Figure 2-4. Strip mall energy use for three stores in Chicago.

Table 2-1. Energy Goals in the Context of the Design Phase

A typical "integrated" design process includes the following steps in sequence, with energy-related actions flagged (✳).

Activities	Responsibilities	Where to Find Information
1. Select Team 　a. Design Team 　b. QA Provider 　c. Construction Team	Owner evaluates potential service providers and selects team.	Chapter 5, QA1 and QA2
2. Owner's Project Requirements (OPRs)[a] ✳ 　a. Choose Recommendation Table Items 　b. Codes/Standards Requirements	Owner and CxA define the OPRs and goals.	Chapter 2, Table 2-5[b] Chapter 3
3. Define Budget 　a. Develop and Review Design Budget 　b. Develop and Review Construction Budget 　c. Develop and Review QA Budget	Owner, GC, Designer Owner, Designer Owner, GC, Estimator Owner, CxA	Chapter 5, QA4
4. Select Implementation Recommendations ✳ 　a. Specify System Preferences 　b. Update OPR document 　c. Check for Rebate/Incentive Programs	Owner, Designer, GC	Chapter 3 Chapter 5, QA3
5. Develop Design and Construction Schedule	Owner, GC, Designer	Chapter 5, QA5
6. Design Development ✳ 　a. Develop Building Plans, Sections, and Details Incorporating above Strategies 　b. Develop HVAC Load Calculations 　c. Size HVAC Equipment 　d. Integrate QA Specifications into Project Manual 　e. Specify ENERGY STAR® Appliances	Designer, CxA	Chapter 5
7. Construction Documents 　a. Develop Lighting and Equipment Details 　b. Develop Outdoor Air Management Details	Owner, Designer, GC	Chapter 5, Lighting Chapter 5, Outdoor Air
8. Design Review ✳ — Verify That Project Meets Original Goals	Owner, Designer, CxA, GC	Chapter 3 Chapter 5
9. Perform Final Coordination and Integration of Architectural, Mechanical, and Electrical Systems	Designer	Chapter 3 Chapter 5
10. Perform Final Cost Estimates	GC, Estimator	
11. Review Final Design Documents	Owner, Designer, CxA	Chapter 3 Chapter 5, QA6

a. The OPR document is a written document that details the intent for energy efficiency, measurable performance criteria, sustainability, functional requirements, and the expectations of how the facility will be used and operated. See Chapter 3 for specific recommendations for each of the building components. Lists of implementation examples are provided in Chapter 5.

b. Table 2-5 presents four goals along with specific strategies for achieving energy savings in retail construction. Reducing loads (Goal 1), both internal and external, is the most basic. Matching the capacity of energy-using systems to the reduced loads (Goal 2) is also important. Oversized systems cost more and do not operate at their optimum efficiency. Higher efficiency equipment (Goal 3) will use less energy to meet any given load. Thus, high-efficiency equipment, in systems whose capacity matches peak loads, serving a building designed and constructed to the lowest practical loads, will result in the lowest energy use and cost. And finally, Goal 4 addresses the integration of building systems to increase energy savings potential.

QUALITY ASSURANCE: IN-HOUSE OR THIRD PARTY?

Users of this Guide may debate whether to use in-house staff or outside third parties as the Commissioning Authority (CxA) to perform the quality assurance (QA) tasks in the design, construction, and acceptance phases of the project. A case can be made for either approach depending on project budget, design complexity, capabilities of the design and construction team, and availability of local Cx expertise.

While both approaches can be effective, building owners should insist that the QA tasks be carried out by a party who is independent from the design and construction team. Independent review ensures that "fresh eyes" are applied to energy performance QA.

Where the in-house approach is deemed to be in the best interests of the building owner, the QA tasks are best accomplished by personnel with no direct interest in the project. For example, qualified staff working on other projects could be assigned as disinterested parties to check and verify the work of their colleagues. However, building owners can expect to get the most independent QA review from outside third parties. Indeed, most of the literature on building Cx and energy performance QA recommends or requires independent outside reviews. In either case, building owners should expect to bear the cost of approximately 25–50 professional staff hours to carry out the Suggested Commissioning Scope (see next page) depending on project specifics. Additional information can be found in Chapter 5.

Quality Assurance: During the design process, the design team documents its design assumptions (basis of design) and includes them in the OPR document. A party other than the installing contractor, architect, or engineer of record should review the contract document and verify that it incorporates the OPRs and the associated strategies contained in this Guide before the start of construction. The owner's agent, if qualified, can provide the required review. This review, along with subsequent inspection, testing, and reporting, is referred to as *commissioning*. The Commissioning Process is a quality-oriented process for achieving, verifying, and documenting that the performance of facilities, systems, and assemblies meets defined objectives and criteria.

The reviewer provides the owner and designers with written comments outlining where items do not comply with these defined objectives and criteria selected from this Guide. Comments should be resolved and any required changes should be completed before start of construction. The owner may choose to use an outside third party to perform this review.

Once the Design Phase is completed, the party that is independent of the design and construction team fulfils the QA role to ensure that the goals, strategies, and recommendations are actually installed and achieved. This Guide provides recommendations to ensure that the goals, strategies, and actions selected are properly executed during the later stages of the building life cycle in Chapter 4 under "Quality Assurance."

SUGGESTED COMMISSIONING SCOPE

- Review the OPRs and the designers' basis-of-design documentation for completeness and clarity. The information provided by the design team for review should include project and design goals, measurable performance criteria, budgets, schedules, success criteria, and supporting information.
- Develop project-specific commissioning/quality assurance (Cx/QA) specifications for building envelope and electrical, mechanical, and plumbing systems that will be verified during the delivery of the project. The specifications will incorporate Cx/QA activities into the construction process and provide a clear understanding to all participants of their specific roles, responsibilities, and effort. The Guide specifications will be reviewed, modified, and blended into the construction documents by the designers.
- Conduct one design review of the construction documents before 100% completion. A review before construction document completion (around 90% completion recommended) allows any changes to be incorporated. The review will focus on ensuring the design is consistent with the OPRs and the designers' basis of design and that all construction requirements are clear and well coordinated. It is also intended to ensure that the specifications describe the roles and responsibilities of all parties to the Cx process so that contractors have a clear understanding of their responsibilities. Prepare a report identifying concerns and opportunities, and use it in working with the owner and designers to develop a collaborative partnership that will ensure delivery of a high-quality building that performs as intended. Provide a report that tracks issues to resolution and follow a collaborative process to facilitate resolution.
- Conduct one two-hour meeting to discuss review comments and adjudicate issues with the design team, and issue a final report illustrating the disposition of each issue raised. Use the report to verify during construction site visits that issues were corrected.
- If a pre-bid meeting is held with bidding contractors, participate in it to emphasize the inclusion of Cx and describe the Cx process for the specific project.
- Prepare construction checklists and the Cx/QA plan and conduct a one-hour meeting with the project team reviewing QA procedures, roles, and responsibilities and establishing a tentative schedule of Cx/QA activities. During the meeting provide the construction checklists to the contractors for their use during the delivery process.
- Review submittal information for systems being commissioned and provide appropriate comments to team. Based on the submittal information, develop test procedures that will be used to verify system performance and distribute to the team.
- Conduct two site visits during construction to observe construction techniques and to identify issues that may affect performance. Review issues with appropriate team members at the end of each site visit in accordance with established communication protocols, and issue one report per visit documenting findings. Establish and maintain an issues log for tracking issues identified.
- Direct and witness testing and document results. Issues identified will be documented in the issues log and tracked to resolution. General contractor (GC) will schedule testing activities and ensure that responsible parties needed for verification are present.
- Review O&M information to ensure warranty requirements and preventive maintenance information required are part of the documentation along with a copy of the OPR and basis-of-design information.
- Witness training of O&M staff to help ensure that O&M staff understands the systems and their operation, warranty responsibilities, and preventive maintenance requirements.

2. CONSTRUCTION

The best designs yield the expected energy savings when the construction plans and specifications are correctly designed and executed. This section outlines what actions the project team can perform to assist in meeting the energy goals. (See Table 2-2 to identify the appropriate time in the process to apply specific recommendations from this Guide.)

During construction, the independent CxA conducts site visits to verify the building envelope construction and that the rough-in of the HVAC and electrical systems meet the OPRs. The purposes of these site visits are as follows:

- **Observations for Operability and Maintainability.** Participate in an ongoing review of the building envelope, mechanical systems, and electrical systems. Prepare field notes and deficiency lists and distribute to the owner, designer, and GC.
- **Verify Access Requirements.** Review shop drawings and perform construction observations to verify that the required access to systems and equipment has been provided.
- **Review Test and Balance (TAB) Plan.** Meet with the construction team to review the TAB plan and schedule required.
- **Random Spot Verification.** Randomly verify installation checklists completed by contractors.

A written report on the site visit that documents issues requiring resolution by the design and/or construction team should be provided. The estimated level effort for the CxA's written report is two to four hours during the construction phase for the size of small retail buildings covered by this Guide.

3. ACCEPTANCE

At this final stage of construction, the project team and the independent CxA verify that systems are operating as intended. When the team is satisfied that all systems are performing as intended, the QA effort of the design and construction team is complete. (See Table 2-3 to identify the appropriate time in the process to apply specific recommendations from this Guide.)

4. OCCUPANCY

During the first year of operation, the building owner needs to review the overall operation and performance of the building. Building systems not performing as expected should be discussed with the design and construction team with issues resolved during the warranty period. The CxA may be brought in to help resolve any Cx/warranty issues. (See Table 2-4 to identify the appropriate time in the process to apply specific recommendations from this Guide.)

5. OPERATION

Energy use and additions of energy-consuming equipment need to be documented and compared against previous data to determine if the building and its systems are operating at peak performance for the life of the building. The CxA needs to provide the ongoing method for monitoring the energy consumption of the building.

Reducing the actual energy use of small retail buildings will be enhanced when advisory energy-tracking information is conveyed to the owner or owner's staff as part of the design package. This information should be developed in simple language and format. This will allow the end-user to track and benchmark the facility's utility bills and take corrective action to maintain the intended efficiency of the original design. This ongoing Cx will require some additional cost but will typically save substantially more by preventing efficiency degradation in the facilities' energy systems. Additional information on energy-effective operation and ongoing energy management is available in *ASHRAE Handbook—HVAC Applications*.

Table 2-2. Energy Goals in the Context of the Bidding and Construction Process

Activities	Responsibilities	Where to Find Information

1. Pre-Bid Conference
    Discuss importance of energy systems to contractors/subcontractors
    Define quality control/Cx role
	Owner, Designer, CxA, GC	Chapter 5, QA7

2. Progress Meetings
    Regular updates on energy-efficiency-related measures
    Scheduling/update QA
	Owner, Designer, GC	

3. Envelope/Energy Systems QA
    QA building envelope construction
    QA HVAC systems
    QA lighting systems
	CxA	Chapter 5, QA8 and QA9

Table 2-3. Energy Goals in the Context of the Acceptance Phase

Activities	Responsibilities	Where to Find Information
1. Assemble punch list of required items to be completed	GC	
2. Performance testing, as required of GC and subcontractors	GC, Subcontractors	Chapter 5, QA10
3. Building is identified as substantially complete	Owner, Designer, CxA	Chapter 5, QA11
4. Maintenance manual submitted and accepted	Owner, Designer, CxA	Chapter 5, QA12
5. Resolve quality control issues identified throughout the construction phase	Owner, Designer, CxA	Chapter 5, QA13
6. Final acceptance	Owner, Designer	Chapter 5, QA14

Table 2-4. Energy Goals in the Context of the Occupancy Phase

Activities	Responsibilities	Where to Find Information
1. Establish building maintenance program	Owner and staff, CxA	Chapter 5, QA15
2. Create post-occupancy punch list	Owner and staff	
3. Monitor post-occupancy performance	Owner and CxA	Chapter 5, QA16

ENERGY GOALS AND STRATEGIES

Numerous goals, listed in Table 2-5, provide detailed strategies and recommendations to meet the 30% energy use reduction target. The related how-to information in this Guide identifies selected energy-saving measures to meet major energy design goals.

Table 2-5. Energy Goals and Strategies

Prioritize Goals	General Strategies	Detailed Strategies	Recommendations (see Chapter 3)	How-To Tips (see Chapter 5)
Goal 1. Reduce loads on energy-using systems				
Reduce internal loads	**Equipment and Appliances:** Reduce both cooling loads and energy use	Use more efficient equipment and appliances	Use low-energy computers and monitors; use ENERGY STAR® equipment	PL1–3
		Use controls to minimize usage and waste	Turn off or use "sleep mode" on computers, monitors, copiers, and other equipment	PL3
		Educate building staff	Provide training, brochures, and other material to encourage energy efficiency	EL5, DL1, 2, 9
	Lighting: Reduce both cooling loads and energy use	Maximize the benefits of daylighting	Vertical glazing, skylights, interior lighting	EN22–32, DL1–10, EL15
		Use skylights and north-facing clerestories to daylight interior zones	Skylights and vertical glazing	EN22, 24, DL3–7
		Use efficient electric lighting system	Interior lighting	EL1–EL25
		Use separate controls for lighting in areas near windows	Interior lighting	DL1, 6–9, EL5, 7–8
		Use automatic controls to turn off lights when not in use	Interior lighting	EL13, 14
Reduce heat gain/ loss through building envelope	**Building Envelope:** Control solar gain to reduce cooling load through windows	Use beneficial building form and orientation		EN26, 28
		Minimize windows east and west, maximize north and south	Vertical glazing	EN26, 28
		Use glazing with low solar heat gain coefficient (SHGC)	Vertical glazing, skylights	EN22–24, 29, 30
		External shade glazing to reduce solar heat gain and glare	Vertical glazing	EN26

Table 2-5. Energy Goals and Strategies (*Continued*)

Prioritize Goals	General Strategies	Detailed Strategies	Recommendations (see Chapter 3)	How-To Tips (see Chapter 5)
Reduce heat gain/loss through building envelope (*continued*)	Control solar gain to reduce cooling load through windows	Use vegetation on S/E/W to control solar heat gain (and glare)	Vertical glazing	EN31
	Reduce solar gain through opaque surfaces to reduce cooling load	Increase insulation of opaque surfaces	Roofs, walls, floors, doors	EN2–20
		Increase roof surface reflectance and emittance	Roofs	EN1
		Shade building surfaces with deciduous or coniferous trees as appropriate for surface orientation		
	Reduce conductive heat gain and loss through building envelope	Increase insulation on roof, walls, floor, slabs, and doors and decrease window U-factor	Roofs, walls, floors, doors, vertical glazing	EN2–20
	Reduce air infiltration	Provide continuous air barrier		
	Reduce heat gain or loss from ventilation exhaust air	Use energy recovery to precondition outdoor air	Energy recovery	
Reduce thermal loads	Utilize passive solar designs	Use thermal storage, trombe walls, interior mass		EN30
Reduce HVAC loads	Reduce heat gain and loss in ductwork	Insulate ductwork	HVAC	HV10
		No ductwork outside the building conditioned space	HVAC	HV9
Refine building to suit local conditions	Consider natural ventilation, highest potential in marine climates, high potential in dry climates	Operable windows with screens so that air conditioning and heating are not necessary during transition periods	Vertical glazing	EN27
		For buildings with operable windows, design building layout for effective cross-ventilation	Vertical glazing	EN27

Table 2-5. Energy Goals and Strategies (Continued)

Prioritize Goals	General Strategies	Detailed Strategies	Recommendations (see Chapter 3)	How-To Tips (see Chapter 5)
Goal 2. Size HVAC systems for reduced loads				
Properly size equipment	Calculate load		HVAC	HV3
	Size equipment		HVAC	HV1, 2, 4
Goal 3. Use more efficient systems				
Use more efficient HVAC systems	Select efficient cooling equipment	Meet or exceed listed equipment efficiencies in "Recommendations" chapter	HVAC	HV1, 2, 4, 6, 17
		Meet or exceed listed part-load performances in "Recommendations" chapter	HVAC	HV1, 2, 4, 6, 21
	Select efficient heating equipment	Meet or exceed listed equipment efficiencies in "Recommendations" chapter	HVAC	HV1, 2, 6, 16, 20
	Select efficient energy recovery equipment	Meet or exceed listed equipment efficiencies in "Recommendations" chapter		HV5, 17
Improve outdoor air ventilation	Control outdoor air dampers	Use air economizer		HV7, 14
		Use demand-controlled ventilation	Ventilation	HV7, 14, 22
		Shut off outdoor air and exhaust air dampers during unoccupied periods	Ventilation	HV7, 8, 14
Improve fan power	Design efficient duct distribution system	Minimize duct and fitting losses	Ducts	HV9, 18, 19
	Reduce duct leakage	Seal all duct joints and seams	Ducts	HV11
	Select efficient motors	Use high-efficiency motors		HV12

Table 2-5. Energy Goals and Strategies *(Continued)*

Prioritize Goals	General Strategies	Detailed Strategies	Recommendations (see Chapter 3)	How-To Tips (see Chapter 5)
Improve HVAC controls	Use control strategies that reduce energy use	Divide building into thermal zones	HVAC	HV13, 21
		Use time-of-day scheduling, temperature setback and setup, pre-occupancy purge	HVAC	HV14
Ensure proper air distribution	Test, adjust, and balance the air distribution system	Use industry-accepted procedures	HVAC	HV15
Use more efficient SWH systems	Select efficient SWH equipment	Meet or exceed listed equipment efficiencies in "Recommendations" chapter	SWH	WH1–4
	Minimize distribution losses	Use point-of-use units	SWH	WH1–2
		Minimize pipe distribution	SWH	WH5
		Insulate piping	SWH	WH6
Use more efficient lighting	More efficient interior lighting	Do not use incandescent lighting unless it will be used infrequently. Avoid incandescent sources except for specific task requirements.		EL11
		Use more efficient electric lighting system	More efficient lamps, ballasts, ceiling fixtures, and case lighting	EL1–4, 7–12, 19–25
	More efficient exterior lighting	Use more efficient electric lighting systems	More efficient exterior lighting sources	EL27, EX2
Goal 4. Refine systems integration				
Integrate building systems	Integrate systems—high-efficiency adv. case			EN23, DL8, 10, QA1
	Integrate systems—daylight adv. case		Advanced daylighting option	EN23, DL8, 10, QA1

Recommendations by Climate 3

Users should determine the recommendations for their construction project by first locating the correct climate zone. The U.S. Department of Energy (DOE) has identified eight climate zones for the United States, with each defined by county borders, as shown in Figure 3-1. This Guide uses these DOE climate zones in defining energy recommendations that vary by climate.

This chapter contains a unique set of energy-efficient recommendations for each climate zone. The recommendation tables represent *a way*, but *not the only way*, for reaching the 30% energy savings target over Standard 90.1–1999. Other approaches may also save energy, but they are not part of the scope of this Guide; assurance of those savings is left to the user. The user should note that the recommendation tables do not include all of the components listed in Standard 90.1 since the Guide focuses only on the primary energy systems within a building.

When a recommendation is provided, the recommended value differs from the requirements in Standard 90.1–1999. When "No recommendation" is indicated, the user must meet at least the minimum requirements of Standard 90.1-1999 or the requirements of local codes whenever they exceed the requirements of Standard 90.1-1999.

Each of the climate zone recommendation tables includes a set of common items arranged by building subsystem: envelope, lighting, HVAC, and service water heating (SWH). Recommendations are included for each item, or subsystem, by component within that subsystem. For some subsystems, recommendations depend on the construction type. For example, insulation values are given for mass, metal building, and steel-framed and wood-framed wall types. For other subsystems, recommendations are given for each subsystem attribute. For example, vertical glazing recommendations are given for size, thermal transmittance, solar heat gain coefficient (SHGC), window orientation, and exterior sun control.

BONUS SAVINGS

Chapter 5 provides additional recommendations and strategies for savings for plug loads and exterior lighting over and above the 30% savings recommendations contained in the following eight climate regions.

The fourth column in each table lists references to how-to tips for implementing the recommended criteria. The tips are found in Chapter 5 under separate sections coded for envelope (EN), daylighting (DL), electric lighting (EL), HVAC systems and equipment (HV), and water heating systems and equipment (WH) suggestions. Besides how-to tips for design and maintenance suggestions that represent good practice, these tips include cautions for what to avoid. Important QA considerations and recommendations are also given for the building design, construction, and post-occupancy phases. Note that each tip is tied to the applicable climate zone in Chapter 5. The final column is provided as a simple checklist to identify the recommendations being used for a specific building design and construction.

The recommendations presented are either minimum or maximum values. Minimum values include R-values, mean lumens/watt, SEER, SRI, EER, IPLV, AFUE, E_c, HSPF, COP, E_t, EF, and insulation thicknesses. Maximum values include U-factors, SHGC, area, LPD, and friction rate.

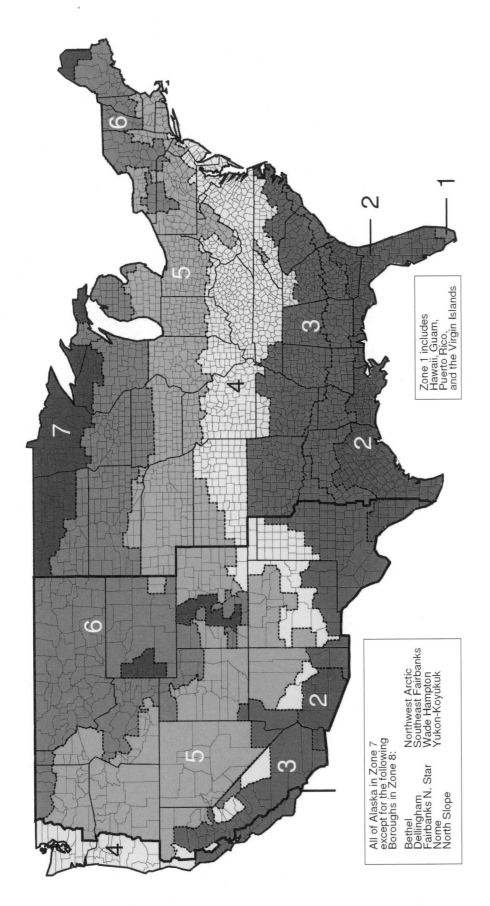

All of Alaska in Zone 7
except for the following
Boroughs in Zone 8:

Bethel   Northwest Arctic
Dellingham   Southeast Fairbanks
Fairbanks N. Star Wade Hampton
Nome   Yukon-Koyukuk
North Slope

Zone 1 includes
Hawaii, Guam,
Puerto Rico,
and the Virgin Islands

Figure 3-1. DOE climate zone map. A list of counties and their respective climate zones can be found on the following pages and at www.energycodes.gov.

Zone 1

Florida

Broward
Miami-Dade
Monroe

Guam

Hawaii

Puerto Rico

U.S. Virgin Islands

Climate Zone 1 Recommendation Table for Small Retail Buildings

Item	Component	Recommendation (Minimum or Maximum)	How-To Tips in Chapter 5	✓
Envelope — Roof	Insulation entirely above deck	R-15 c.i.	EN1-2, 17, 20-21	
	Metal building	R-19	EN1, 3, 17, 20-21	
	Attic and other	R-30	EN4, 17-18, 20-21	
	Single rafter	R-30	EN5, 17, 20-21	
	Solar reflectance index (SRI)	78	EN1	
Walls	Mass (HC > 7 Btu/ft^2)	No recommendation	EN6, 17, 20-21	
	Metal building	R-13	EN7, 17, 20-21	
	Steel framed	R-13	EN8, 17, 20-21	
	Wood framed and other	R-13	EN9, 17, 20-21	
	Below-grade walls	No recommendation	EN10, 17, 20-21	
Floors	Mass	R-4.2 c.i.	EN11, 17, 20-21	
	Steel framed	R-19	EN12, 17, 20-21	
	Wood framed and other	R-19	EN12, 17, 20-21	
Slabs	Unheated	No recommendation	EN13, 17, 19-21	
	Heated	R-7.5 for 12 in.	EN14, 17, 19-21	
Doors— Opaque	Swinging	U-0.70	EN15, 20-21	
	Non-swinging	U-1.45	EN16, 20-21	
Vertical Glazing Including Doors	Area (percent of gross wall)	40%	EN22-23, 27, 28, 29	
	Thermal transmittance	U-0.69	EN22, 25	
	Solar heat gain coefficient (SHGC)	N, S, E, W - 0.44; N only—0.44	EN22	
	Exterior sun control (S, E, W only)	Projection factor > 0.5	EN26, DL3	
Skylights	Area (percent of gross roof)	3%	EN24	
	Thermal transmittance	U-1.36		
	Solar heat gain coefficient (SHGC)	0.19	DL3-10	
Lighting — Interior Lighting	Lighting power density (LPD)	1.3 W/ft^2	EL1, 3, 4, 14, 16-25	
	Linear fluorescent with high-performance electronic ballast	91 mean lm/W	EL7, 8	
	All other sources	50 mean lm/W	EL9,10	
	Dimming controls for daylight harvesting under skylights	Dim fixtures within 10 ft of skylight edge	DL1-9, EL15	
	Occupancy controls	Auto-off all non-sales rooms	EL13	
	Interior room surface reflectances in locations with daylighting	80%+ on ceilings, 70%+ on walls	DL2, EL5	
Additional Interior Lighting for Sales Floor	Additional LPD for adjustable lighting equipment that is specifically designed and directed to highlight merchandise and is automatically controlled separately from the general lighting	0.4 W/ft^2 (spaces not listed below) 0.6 W/ft^2 (sporting goods, small electronics) 0.9 W/ft^2 (furniture, clothing, cosmetics, and artwork) 1.5 W/ft^2 (jewelry, crystal, china)	EL1, 2, 10, 11, 14, 20, 21	
	Sources	Halogen IR or CMH	EL10-12	
Exterior Lighting	Façade and externally illuminated signage lighting	0.2 W/ft^2	EX2, 3, EL5, EL7-12	
HVAC — HVAC	Air conditioner (0-65 kBtuh)	13.0 SEER	HV1-4, 6, 12, 16-17, 20	
	Air conditioner (>65-135 kBtuh)	11.3 EER/11.5 IPLV	HV1-4, 6, 12, 16-17, 20	
	Air conditioner (>135-240 kBtuh)	11.0 EER/11.5 IPLV	HV1-4, 6, 12, 16-17, 20	
	Air conditioner (>240 kBtuh)	10.6 EER/11.2 IPLV	HV1-4, 6, 12, 16-17, 20	
	Gas furnace (0-225 kBtuh - SP)	80% AFUE or E_t	HV1-3, 6, 16, 20	
	Gas furnace (0-225 kBtuh - Split)	80% AFUE or E_t	HV1-3, 6, 16, 20	
	Gas furnace (>225 kBtuh)	80% E_c	HV1-3, 6, 16, 20	
	Heat pump (0-65 kBtuh)	13.0 SEER/7.7 HSPF	HV1-4, 6, 12, 16-17, 20	
	Heat pump (>65-135 kBtuh)	10.6 EER/11.0 IPLV/3.2 COP	HV1-4, 6, 12, 16-17, 20	
	Heat pump (>135 kBtuh)	10.1 EER/11.5 IPLV/3.1 COP	HV1-4, 6, 12, 16-17, 20	
Economizer	Air conditioners & heat pumps- SP	No recommendation	HV23	
Ventilation	Outdoor air damper	Motorized control	HV7-8	
	Demand control	CO_2 sensors	HV7, 22	
Ducts	Friction rate	0.08 in. w.c./100 ft	HV9, 18	
	Sealing	Seal class B	HV11	
	Location	Interior only	HV9	
	Insulation level	R-6	HV10	
SWH — Service Water Heating	Gas storage (> 75 kBtuh)	90% E_t	WH1-4	
	Gas Instantaneous	0.81 EF or 81% E_t	WH1-4	
	Electric storage (≤ 12 kW and > 20 gal)	EF > 0.99 – 0.0012xVolume	WH1-4	
	Pipe insulation (d < 1½ in./ d ≥ 1½ in.)	1 in./ 1½ in.	WH6	

Note: If the table contains "No recommendation" for a component, the user must meet the more stringent of either Standard 90.1 or the local code requirements in order to reach the 30% savings target.

Zone 2

Alabama	Levy	Liberty	West Feliciana	Houston
	Liberty	Long		Jackson
Baldwin	Madison	Lowndes	**Mississippi**	Jasper
Mobile	Manatee	McIntosh		Jefferson
	Marion	Miller	Hancock	Jim Hogg
Arizona	Martin	Mitchell	Harrison	Jim Wells
	Nassau	Pierce	Jackson	Karnes
La Paz	Okaloosa	Seminole	Pearl River	Kenedy
Maricopa	Okeechobee	Tattnall	Stone	Kinney
Pima	Orange	Thomas		Kleberg
Pinal	Osceola	Toombs	**Texas**	La Salle
Yuma	Palm Beach	Ware		Lavaca
	Pasco	Wayne	Anderson	Lee
California	Pinellas		Angelina	Leon
	Polk	**Louisiana**	Aransas	Liberty
Imperial	Putnam		Atascosa	Limestone
	Santa Rosa	Acadia	Austin	Live Oak
Florida	Sarasota	Allen	Bandera	Madison
	Seminole	Ascension	Bastrop	Matagorda
Alachua	St. Johns	Assumption	Bee	Maverick
Baker	St. Lucie	Avoyelles	Bell	McLennan
Bay	Sumter	Beauregard	Bexar	McMullen
Bradford	Suwannee	Calcasieu	Bosque	Medina
Brevard	Taylor	Cameron	Brazoria	Milam
Calhoun	Union	East Baton	Brazos	Montgomery
Charlotte	Volusia	Rouge	Brooks	Newton
Citrus	Wakulla	East Feliciana	Burleson	Nueces
Clay	Walton	Evangeline	Caldwell	Orange
Collier	Washington	Iberia	Calhoun	Polk
Columbia		Iberville	Cameron	Real
DeSoto	**Georgia**	Jefferson	Chambers	Refugio
Dixie		Jefferson Davis	Cherokee	Robertson
Duval	Appling	Lafayette	Colorado	San Jacinto
Escambia	Atkinson	Lafourche	Comal	San Patricio
Flagler	Bacon	Livingston	Coryell	Starr
Franklin	Baker	Orleans	DeWitt	Travis
Gadsden	Berrien	Plaquemines	Dimmit	Trinity
Gilchrist	Brantley	Pointe Coupee	Duval	Tyler
Glades	Brooks	Rapides	Edwards	Uvalde
Gulf	Bryan	St. Bernard	Falls	Val Verde
Hamilton	Camden	St. Charles	Fayette	Victoria
Hardee	Charlton	St. Helena	Fort Bend	Walker
Hendry	Chatham	St. James	Freestone	Waller
Hernando	Clinch	St. John the Baptist	Frio	Washington
Highlands	Colquitt	St. Landry	Galveston	Webb
Hillsborough	Cook	St. Martin	Goliad	Wharton
Holmes	Decatur	St. Mary	Gonzales	Willacy
Indian River	Echols	St. Tammany	Grimes	Williamson
Jackson	Effingham	Tangipahoa	Guadalupe	Wilson
Jefferson	Evans	Terrebonne	Hardin	Zapata
Lafayette	Glynn	Vermilion	Harris	Zavala
Lake	Grady	Washington	Hays	
Lee	Jeff Davis	West Baton	Hidalgo	
Leon	Lanier	Rouge	Hill	

Climate Zone 2 Recommendation Table for Small Retail Buildings

	Item	Component	Recommendation (Minimum or Maximum)	How-To Tips in Chapter 5	✓
Envelope	Roof	Insulation entirely above deck	R-15 c.i.	EN1-2, 17, 20-21	
		Metal building	R-19	EN1, 3, 17, 20-21	
		Attic and other	R-38	EN4, 17-18, 20-21	
		Single rafter	R-38	EN5, 17, 20-21	
		Solar reflectance index (SRI)	78	EN1	
	Walls	Mass (HC > 7 Btu/ft^2)	R-7.6 c.i.	EN6, 17, 20-21	
		Metal building	R-13	EN7, 17, 20-21	
		Steel framed	R-13	EN8, 17, 20-21	
		Wood framed and other	R-13	EN9, 17, 20-21	
		Below-grade walls	No recommendation	EN10, 17, 20-21	
	Floors	Mass	R-6.3 c.i.	EN11, 17, 20-21	
		Steel framed	R-19	EN12, 17, 20-21	
		Wood framed and other	R-19	EN12, 17, 20-21	
	Slabs	Unheated	No recommendation	EN13, 17, 19-21	
		Heated	R-7.5 for 12 in.	EN14, 17, 19-21	
	Doors— Opaque	Swinging	U-0.70	EN15, 20-21	
		Non-swinging	U-1.45	EN16, 20-21	
	Vertical Glazing Including Doors	Area (percent of gross wall)	40%	EN22-23, 27, 28, 29	
		Thermal transmittance	U-0.49	EN22, 25	
		Solar heat gain coefficient (SHGC)	N, S, E, W - 0.40; N only— 0.40	EN22	
		Exterior sun control (S, E, W only)	Projection factor > 0.5	EN26, DL3	
	Skylights	Area (percent of gross roof)	3%	EN24	
		Thermal transmittance	U-1.36		
		Solar heat gain coefficient (SHGC)	0.19	DL3-10	
Lighting	Interior Lighting	Lighting power density (LPD)	1.3 W/ft^2	EL1, 3, 4, 14, 16-25	
		Linear fluorescent with high-performance electronic ballast	91 mean lm/W	EL7, 8	
		All other sources	50 mean lm/W	EL9,10	
		Dimming controls for daylight harvesting under skylights	Dim fixtures within 10 ft of skylight edge	DL1-9, EL15	
		Occupancy controls	Auto-off all non-sales rooms	EL13	
		Interior room surface reflectances in locations with daylighting	80%+ on ceilings, 70%+ on walls	DL2, EL5	
	Additional Interior Lighting for Sales Floor	Additional LPD for adjustable lighting equipment that is specifically designed and directed to highlight merchandise and is automatically controlled separately from the general lighting	0.4 W/ft^2 (spaces not listed below) 0.6 W/ft^2 (sporting goods, small electronics) 0.9 W/ft^2 (furniture, clothing, cosmetics, and artwork) 1.5 W/ft^2 (jewelry, crystal, china)	EL1, 2, 10, 11, 14, 20, 21	
		Sources	Halogen IR or CMH	EL10-12	
	Exterior Lighting	Façade and externally illuminated signage lighting	0.2 W/ft^2	EX2, 3, EL5, EL7-12	
HVAC	HVAC	Air conditioner (0-65 kBtuh)	13.0 SEER	HV1-4, 6, 12, 16-17, 20	
		Air conditioner (>65-135 kBtuh)	11.3 EER/11.5 IPLV	HV1-4, 6, 12, 16-17, 20	
		Air conditioner (>135-240 kBtuh)	11.0 EER/11.5 IPLV	HV1-4, 6, 12, 16-17, 20	
		Air conditioner (>240 kBtuh)	10.6 EER/11.2 IPLV	HV1-4, 6, 12, 16-17, 20	
		Gas furnace (0-225 kBtuh - SP)	80% AFUE or E_t	HV1-2, 6, 16, 20	
		Gas furnace (0-225 kBtuh - Split)	80% AFUE or E_t	HV1-2, 6, 16, 20	
		Gas furnace (>225 kBtuh)	80% E_c	HV1-2, 6, 16, 20	
		Heat pump (0-65 kBtuh)	13.0 SEER/7.7 HSPF	HV1-4, 6, 12, 16-17, 20	
		Heat pump (>65-135 kBtuh)	10.6 EER/11.0 IPLV/3.2 COP	HV1-4, 6, 12, 16-17, 20	
		Heat pump (>135 kBtuh)	10.1 EER/11.5 IPLV/3.1 COP	HV1-4, 6, 12, 16-17, 20	
	Economizer	Air conditioners & heat pumps- SP	No recommendation	HV23	
	Ventilation	Outdoor air damper	Motorized control	HV7-8	
		Demand control	CO_2 sensors	HV7, 22	
	Ducts	Friction rate	0.08 in. w.c./100 ft	HV9, 18	
		Sealing	Seal class B	HV11	
		Location	Interior only	HV9	
		Insulation level	R-6	HV10	
SWH	Service Water Heating	Gas storage (> 75 kBtuh)	90% E_t	WH1-4	
		Gas Instantaneous	0.81 EF or 81% E_t	WH1-4	
		Electric storage (≤ 12 kW and > 20 gal)	EF > 0.99 – 0.0012xVolume	WH1-4	
		Pipe insulation (d < 1½ in./ d ≥ 1½ in.)	1 in./ 1½ in.	WH6	

Note: If the table contains "No recommendation" for a component, the user must meet the more stringent of either Standard 90.1 or the local code requirements in order to reach the 30% savings target.

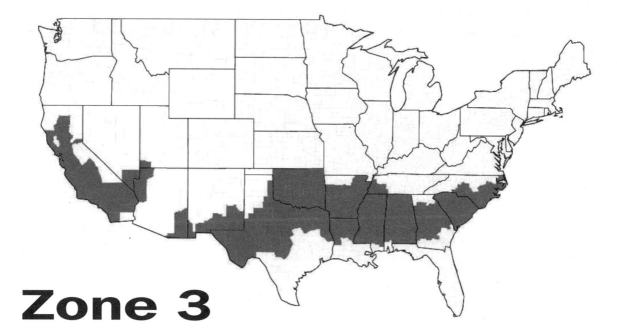

Zone 3

Alabama

All counties except:
Baldwin
Mobile

Arizona

Cochise
Graham
Greenlee
Mohave
Santa Cruz

Arkansas

All counties except:
Baxter
Benton
Boone
Carroll
Fulton
Izard
Madison
Marion
Newton
Searcy
Stone
Washington

California

All counties except:
Alpine
Amador
Calaveras
Del Norte
El Dorado
Humboldt
Imperial
Inyo
Lake
Lassen
Mariposa
Modoc
Mono
Nevada
Plumas
Sierra
Siskiyou
Trinity
Tuolumne

Georgia

All counties except:
Appling
Atkinson
Bacon
Baker
Banks
Berrien
Brantley
Brooks
Bryan
Catoosa
Camden
Charlton
Chatham
Chattooga
Clinch
Colquitt
Cook
Dade
Dawson
Decatur
Echols
Effingham
Evans
Fannin
Floyd
Franklin
Gilmer
Glynn
Gordon
Grady
Habersham
Hall
Jeff Davis
Lanier
Liberty
Long
Lowndes
Lumpkin
McIntosh
Miller
Mitchell
Murray
Pickens
Pierce
Rabun
Seminole
Stephens
Tattnall
Thomas
Toombs
Towns

Union
Walker
Ware
Wayne
White
Whitfield

Louisiana

Bienville
Bossier
Caddo
Caldwell
Catahoula
Claiborne
Concordia
De Soto
East Carroll
Franklin
Grant
Jackson
La Salle
Lincoln
Madison
Morehouse
Natchitoches
Ouachita
Red River
Richland
Sabine
Tensas
Union
Vernon
Webster
West Carroll
Winn

Mississippi

All counties except:
Hancock
Harrison
Jackson
Pearl River
Stone

New Mexico

Chaves
Dona Ana
Eddy
Hidalgo
Lea
Luna
Otero

Nevada

Clark

Texas

Andrews
Archer
Baylor
Blanco
Borden
Bowie
Brewster
Brown
Burnet
Callahan
Camp
Cass
Childress
Clay
Coke
Coleman
Collingsworth
Collin
Comanche
Concho
Cottle
Cooke
Crane
Crockett
Crosby
Culberson
Dallas
Dawson
Delta
Denton
Dickens
Eastland
Ector
El Paso
Ellis
Erath
Fannin
Fisher
Foard
Franklin
Gaines
Garza
Gillespie
Glasscock
Grayson
Gregg
Hall
Hamilton
Hardeman

Harrison
Haskell
Hemphill
Henderson
Hood
Hopkins
Howard
Hudspeth
Hunt
Irion
Jack
Jeff Davis
Johnson
Jones
Kaufman
Kendall
Kent
Kerr
Kimble
King
Knox
Lamar
Lampasas
Llano
Loving
Lubbock
Lynn
Marion
Martin
Mason
McCulloch
Menard
Midland
Mills
Mitchell
Montague
Morris
Motley
Nacogdoches
Navarro
Nolan
Palo Pinto
Panola
Parker
Pecos
Presidio
Rains
Reagan
Reeves
Red River
Rockwall
Runnels
Rusk
Sabine
San Augustine

San Saba
Schleicher
Scurry
Shackelford
Shelby
Smith
Somervell
Stephens
Sterling
Stonewall
Sutton
Tarrant
Taylor
Terrell
Terry
Throckmorton
Titus
Tom Green
Upshur
Upton
Van Zandt
Ward
Wheeler
Wichita
Wilbarger
Winkler
Wise
Wood
Young

Utah

Washington

North Carolina

Anson
Beaufort
Bladen
Brunswick
Cabarrus
Camden
Carteret
Chowan
Columbus
Craven
Cumberland
Currituck
Dare
Davidson
Duplin
Edgecombe
Gaston
Greene
Hoke

Hyde
Johnston
Jones
Lenoir
Martin
Mecklenburg
Montgomery
Moore
New Hanover
Onslow
Pamlico
Pasquotank
Pender
Perquimans
Pitt
Randolph
Richmond
Robeson
Rowan
Sampson
Scotland
Stanly
Tyrrell
Union
Washington
Wayne
Wilson

Oklahoma

All counties except:
Beaver
Cimarron
Texas

South Carolina

All counties

Tennessee

Chester
Crockett
Dyer
Fayette
Hardeman
Hardin
Haywood
Henderson
Lake
Lauderdale
Madison
McNairy
Shelby
Tipton

Climate Zone 3 Recommendation Table for Small Retail Buildings

	Item	Component	Recommendation (Minimum or Maximum)	How-To Tips in Chapter 5	✓
Envelope	Roof	Insulation entirely above deck	R-20 c.i.	EN1-2, 17, 20-21	
		Metal building	R-13 + R-19	EN1, 3, 17, 20-21	
		Attic and other	R-38	EN4, 17-18, 20-21	
		Single rafter	R-38 + R-5 c.i.	EN5, 17, 20-21	
		Solar reflectance index (SRI)	78	EN1	
	Walls	Mass (HC > 7 Btu/ft^2)	R-11.4 c.i.	EN6, 17, 20-21	
		Metal building	R-13 + R-13	EN7, 17, 20-21	
		Steel framed	R-13 + R-7.5 c.i.	EN8, 17, 20-21	
		Wood framed and other	R-13	EN9, 17, 20-21	
		Below-grade walls	No recommendation	EN10, 17, 20-21	
	Floors	Mass	R-10.4 c.i.	EN11, 17, 20-21	
		Steel framed	R-30	EN12, 17, 20-21	
		Wood framed and other	R-30	EN12, 17, 20-21	
	Slabs	Unheated	No recommendation	EN13, 17, 19-21	
		Heated	R-7.5 for 12 in.	EN14, 17, 19-21	
	Doors - Opaque	Swinging	U-0.70	EN15, 20-21	
		Non-swinging	U-0.50	EN16, 20-21	
	Vertical Glazing Including Doors	Area (percent of gross wall)	40%	EN22-23, 27, 28, 29	
		Thermal transmittance	U-0.41	EN22, 25	
		Solar heat gain coefficient (SHGC)	N, S, E, W - 0.41; N only—0.41	EN22	
		Exterior sun control (S, E, W only)	Projection factor > 0.5	EN26, DL3	
	Skylights	Area (percent of gross roof)	3%	EN24	
		Thermal transmittance	U-0.69		
		Solar heat gain coefficient (SHGC)	0.16	DL3-10	
Lighting	Interior Lighting	Lighting power density (LPD)	1.3 W/ft^2	EL1, 3, 4, 14, 16-25	
		Linear fluorescent with high-performance electronic ballast	91 mean lm/W	EL7, 8	
		All other sources	50 mean lm/W	EL9,10	
		Dimming controls for daylight harvesting under skylights	Dim fixtures within 10 ft of skylight edge	DL1-9, EL15	
		Occupancy controls	Auto-off all non-sales rooms	EL13	
		Interior room surface reflectances in locations with daylighting	80%+ on ceilings, 70%+ on walls	DL2, EL5	
	Additional Interior Lighting for Sales Floor	Additional LPD for adjustable lighting equipment that is specifically designed and directed to highlight merchandise and is automatically controlled separately from the general lighting	0.4 W/ft^2 (spaces not listed below) 0.6 W/ft^2 (sporting goods, small electronics) 0.9 W/ft^2 (furniture, clothing, cosmetics, and artwork) 1.5 W/ft^2 (jewelry, crystal, china)	EL1, 2, 10, 11, 14, 20, 21	
		Sources	Halogen IR or CMH	EL10-12	
	Exterior Lighting	Façade and externally illuminated signage lighting	0.2 W/ft^2	EX2, 3, EL5, EL7-12	
HVAC	HVAC	Air conditioner (0-65 kBtuh)	13.0 SEER	HV1-4, 6, 12, 16-17, 20	
		Air conditioner (>65-135 kBtuh)	11.0 EER/11.4 IPLV	HV1-4, 6, 12, 16-17, 20	
		Air conditioner (>135-240 kBtuh)	10.8 EER/11.2 IPLV	HV1-4, 6, 12, 16-17, 20	
		Air conditioner (>240 kBtuh)	10.0 EER/10.4 IPLV	HV1-4, 6, 12, 16-17, 20	
		Gas furnace (0-225 kBtuh - SP)	80% AFUE or E$_t$	HV1-2, 6, 16, 20	
		Gas furnace (0-225 kBtuh - Split)	80% AFUE or E$_t$	HV1-2, 6, 16, 20	
		Gas furnace (>225 kBtuh)	80% E$_c$	HV1-2, 6, 16, 20	
		Heat pump (0-65 kBtuh)	13.0 SEER/7.7 HSPF	HV1-4, 6, 12, 16-17, 20	
		Heat pump (>65-135 kBtuh)	10.6 EER/11.0 IPLV/3.2 COP	HV1-4, 6, 12, 16-17, 20	
		Heat pump (>135 kBtuh)	10.1 EER/11.0 IPLV/3.1 COP	HV1-4, 6, 12, 16-17, 20	
	Economizer	Air conditioners & heat pumps- SP	Cooling capacity > 54 kBtuh	HV23	
	Ventilation	Outdoor air damper	Motorized control	HV7-8	
		Demand control	CO_2 sensors	HV7, 22	
	Ducts	Friction rate	0.08 in. w.c./100 ft	HV9, 18	
		Sealing	Seal class B	HV11	
		Location	Interior only	HV9	
		Insulation level	R-6	HV10	
SWH	Service Water Heating	Gas storage (> 75 kBtuh)	90% E$_t$	WH1-4	
		Gas Instantaneous	0.81 EF or 81% E$_t$	WH1-4	
		Electric storage (≤ 12 kW and > 20 gal)	EF > 0.99 – 0.0012xVolume	WH1-4	
		Pipe insulation (d < 1½ in./ d ≥ 1½ in.)	1 in./ 1½ in.	WH6	

Note: If the table contains "No recommendation" for a component, the user must meet the more stringent of either Standard 90.1 or the local code requirements in order to reach the 30% savings target.

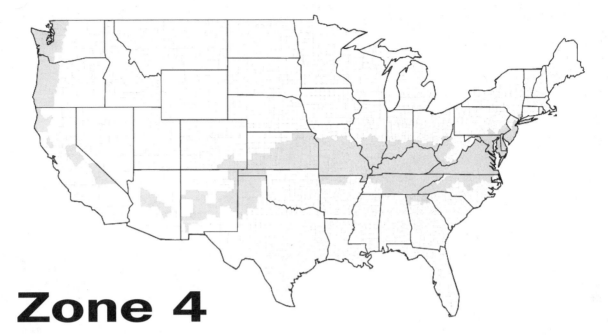

Zone 4

Arizona
Gila
Yavapai

Arkansas
Baxter
Benton
Boone
Carroll
Fulton
Izard
Madison
Marion
Newton
Searcy
Stone
Washington

California
Amador
Calaveras
Del Norte
El Dorado
Inyo
Lake
Mariposa
Trinity
Tuolumne

Colorado
Baca
Las Animas
Otero

Delaware
All counties

District of Columbia

Georgia
Banks
Catoosa
Chattooga
Dade
Dawson
Fannin
Floyd
Franklin
Gilmer
Gordon
Habersham
Hall
Lumpkin
Murray
Pickens
Rabun
Stephens
Towns
Union
Walker
White
Whitfield

Illinois
Alexander
Bond
Brown
Christian
Clay
Clinton
Crawford
Edwards
Effingham
Fayette
Franklin
Gallatin
Hamilton
Hardin
Jackson
Jasper
Jefferson
Johnson
Lawrence
Macoupin
Madison
Marion
Massac
Monroe
Montgomery
Perry
Pope
Pulaski
Randolph
Richland
Saline
Shelby
St. Clair
Union
Wabash
Washington
Wayne
White
Williamson

Indiana
Clark
Crawford
Daviess
Dearborn
Dubois
Floyd
Gibson
Greene
Harrison
Jackson
Jefferson
Jennings
Knox
Lawrence
Martin
Monroe
Ohio
Orange
Perry
Pike
Posey

Ripley
Scott
Spencer
Sullivan
Switzerland
Vanderburgh
Warrick
Washington

Kansas
All counties except:
Cheyenne
Cloud
Decatur
Ellis
Gove
Graham
Greeley
Hamilton
Jewell
Lane
Logan
Mitchell
Ness
Norton
Osborne
Phillips
Rawlins
Republic
Rooks
Scott
Sheridan
Sherman
Smith
Thomas
Trego
Wallace
Wichita

Kentucky
All counties

Maryland
All counties except:
Garrett

Missouri
All counties except:
Adair
Andrew
Atchison
Buchanan
Caldwell
Chariton
Clark
Clinton
Daviess
DeKalb
Gentry
Grundy

Harrison
Holt
Knox
Lewis
Linn
Livingston
Macon
Marion
Mercer
Nodaway
Pike
Putnam
Ralls
Schuyler
Scotland
Shelby
Sullivan
Worth

New Jersey
All counties except:
Bergen
Hunterdon
Mercer
Morris
Passaic
Somerset
Sussex
Warren

New Mexico
Bernalillo
Cibola
Curry
DeBaca
Grant
Guadalupe
Lincoln
Quay
Roosevelt
Sierra
Socorro
Union
Valencia

New York
Bronx
Kings
Nassau
New York
Queens
Richmond
Suffolk
Westchester

North Carolina
Alamance
Alexander
Bertie
Buncombe
Burke

Caldwell
Caswell
Catawba
Chatham
Cherokee
Clay
Cleveland
Davie
Durham
Forsyth
Franklin
Gates
Graham
Granville
Guilford
Halifax
Harnett
Haywood
Henderson
Hertford
Iredell
Jackson
Lee
Lincoln
Macon
Madison
McDowell
Nash
Northampton
Orange
Person
Polk
Rockingham
Rutherford
Stokes
Surry
Swain
Transylvania
Vance
Wake
Warren
Wilkes
Yadkin

Ohio
Adams
Brown
Clermont
Gallia
Hamilton
Lawrence
Pike
Scioto
Washington

Oklahoma
Beaver
Cimarron
Texas

Oregon
Benton
Clackamas

Clatsop
Columbia
Coos
Curry
Douglas
Jackson
Josephine
Lane
Lincoln
Linn
Marion
Multnomah
Polk
Tillamook
Washington
Yamhill

Pennsylvania
Bucks
Chester
Delaware
Montgomery
Philadelphia
York

Tennessee
All counties except:
Chester
Crockett
Dyer
Fayette
Hardeman
Hardin
Haywood
Henderson
Lake
Lauderdale
Madison
McNairy
Shelby
Tipton

Texas
Armstrong
Bailey
Briscoe
Carson
Castro
Cochran
Dallam
Deaf Smith
Donley
Floyd
Gray
Hale
Hansford
Hartley
Hockley
Hutchinson
Lamb
Lipscomb
Moore

Ochiltree
Oldham
Parmer
Potter
Randall
Roberts
Sherman
Swisher
Yoakum

Virginia
All counties

Washington
Clallam
Clark
Cowlitz
Grays Harbor
Island
Jefferson
King
Kitsap
Lewis
Mason
Pacific
Pierce
San Juan
Skagit
Snohomish
Thurston
Wahkiakum
Whatcom

West Virginia
Berkeley
Boone
Braxton
Cabell
Calhoun
Clay
Gilmer
Jackson
Jefferson
Kanawha
Lincoln
Logan
Mason
McDowell
Mercer
Mingo
Monroe
Morgan
Pleasants
Putnam
Ritchie
Roane
Tyler
Wayne
Wirt
Wood
Wyoming

Climate Zone 4 Recommendation Table for Small Retail Buildings

	Item	Component	Recommendation (Minimum or Maximum)	How-To Tips in Chapter 5	✓
Envelope	Roof	Insulation entirely above deck	R-20 c.i.	EN1-2, 17, 20-21	
		Metal building	R-13 + R-19	EN1, 3, 17, 20-21	
		Attic and other	R-38	EN4, 17-18, 20-21	
		Single rafter	R-38 + R-5 c.i.	EN5, 17, 20-21	
		Solar reflectance index (SRI)	No recommendation	EN1	
	Walls	Mass (HC > 7 Btu/ft^2)	R-13.3 c.i.	EN6, 17, 20-21	
		Metal building	R-13 + R-13	EN7, 17, 20-21	
		Steel framed	R-13 + R-7.5 c.i.	EN8, 17, 20-21	
		Wood framed and other	R-13 + R-3.8 c.i.	EN9, 17, 20-21	
		Below-grade walls	R-7.5 c.i.	EN10, 17, 20-21	
	Floors	Mass	R-12.5 c.i.	EN11, 17, 20-21	
		Steel framed	R-30	EN12, 17, 20-21	
		Wood framed and other	R-30	EN12, 17, 20-21	
	Slabs	Unheated	No recommendation	EN13, 17, 19-21	
		Heated	R-7.5 for 12 in.	EN14, 17, 19-21	
	Doors – Opaque	Swinging	U-0.50	EN15, 20-21	
		Non-swinging	U-0.50	EN16, 20-21	
	Vertical Glazing Including Doors	Area (percent of gross wall)	40%	EN22-23, 27, 28, 29	
		Thermal transmittance	U-0.38	EN22, 25	
		Solar heat gain coefficient (SHGC)	N, S, E, W - 0.41; N only—0.41	EN22	
		Exterior sun control (S, E, W only)	Projection factor > 0.5	EN26, DL3	
	Skylights	Area (percent of gross roof)	3%	EN24	
		Thermal transmittance	U-0.69		
		Solar heat gain coefficient (SHGC)	0.32	DL3-10	
Lighting	Interior Lighting	Lighting power density (LPD)	1.3 W/ft^2	EL1, 3, 4, 14, 16-25	
		Linear fluorescent with high-performance electronic ballast	91 mean lm/W	EL7, 8	
		All other sources	50 mean lm/W	EL9,10	
		Dimming controls for daylight harvesting under skylights	Dim fixtures within 10 ft of skylight edge	DL1-9, EL15	
		Occupancy controls	Auto-off all non-sales rooms	EL13	
		Interior room surface reflectances in locations with daylighting	80%+ on ceilings, 70%+ on walls	DL2, EL5	
	Additional Interior Lighting for Sales Floor	Additional LPD for adjustable lighting equipment that is specifically designed and directed to highlight merchandise and is automatically controlled separately from the general lighting	0.4 W/ft^2 (spaces not listed below) 0.6 W/ft^2 (sporting goods, small electronics) 0.9 W/ft^2 (furniture, clothing, cosmetics, and artwork) 1.5 W/ft^2 (jewelry, crystal, china)	EL1, 2, 10, 11, 14, 20, 21	
		Sources	Halogen IR or CMH	EL10-12	
	Exterior	Façade and externally illuminated signage	0.2 W/ft^2	EX2, 3, EL5, EL7-12	
HVAC	HVAC	Air conditioner (0-65 kBtuh)	13.0 SEER	HV1-4, 6, 12, 16-17, 20	
		Air conditioner (>65-135 kBtuh)	11.0 EER/11.4 IPLV	HV1-4, 6, 12, 16-17, 20	
		Air conditioner (>135-240 kBtuh)	10.8 EER/11.2 IPLV	HV1-4, 6, 12, 16-17, 20	
		Air conditioner (>240 kBtuh)	10.0 EER/10.4 IPLV	HV1-4, 6, 12, 16-17, 20	
		Gas furnace (0-225 kBtuh - SP)	80% AFUE or E_t	HV1-2, 6, 16, 20	
		Gas furnace (0-225 kBtuh - Split)	80% AFUE or E_t	HV1-2, 6, 16, 20	
		Gas furnace (>225 kBtuh)	80% E_c	HV1-2, 6, 16, 20	
		Heat pump (0-65 kBtuh)	13.0 SEER/7.7 HSPF	HV1-4, 6, 12, 16-17, 20	
		Heat pump (>65-135 kBtuh)	10.6 EER/11.0 IPLV/3.2 COP	HV1-4, 6, 12, 16-17, 20	
		Heat pump (>135 kBtuh)	10.1 EER/11.0 IPLV/3.1 COP	HV1-4, 6, 12, 16-17, 20	
	Economizer	Air conditioners & heat pumps- SP	Cooling capacity > 54 kBtuh	HV23	
	Ventilation	Outdoor air damper	Motorized control	HV7-8	
		Demand control	CO_2 sensors	HV7, 22	
	Ducts	Friction rate	0.08 in. w.c./100 ft	HV9, 18	
		Sealing	Seal class B	HV11	
		Location	Interior only	HV9	
		Insulation level	R-6	HV10	
SWH	Service Water Heating	Gas storage (> 75 kBtuh)	90% E_t	WH1-4	
		Gas Instantaneous	0.81 EF or 81% E_t	WH1-4	
		Electric storage (≤ 12 kW and > 20 gal)	EF > 0.99 – 0.0012xVolume	WH1-4	
		Pipe insulation (d < 1½ in./ d ≥ 1½ in.)	1 in./ 1½ in.	WH6	

Note: If the table contains "No recommendation" for a component, the user must meet the more stringent of either Standard 90.1 or the local code requirements in order to reach the 30% savings target.

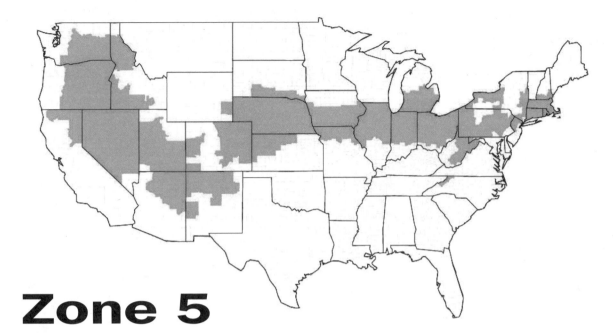

Zone 5

Arizona
Apache
Coconino
Navajo

California
Lassen
Modoc
Nevada
Plumas
Sierra
Siskiyou

Colorado
Adams
Arapahoe
Bent
Boulder
Cheyenne
Crowley
Delta
Denver
Douglas
Elbert
El Paso
Fremont
Garfield
Gilpin
Huerfano
Jefferson
Kiowa
Kit Carson
La Plata
Larimer
Lincoln
Logan
Mesa
Montezuma
Montrose
Morgan
Phillips
Prowers
Pueblo
Sedgwick
Teller
Washington
Weld
Yuma

Connecticut
All counties

Idaho
Ada
Benewah
Canyon
Cassia
Clearwater
Elmore
Gem
Gooding
Idaho
Jerome
Kootenai

Latah
Lewis
Lincoln
Minidoka
Nez Perce
Owyhee
Payette
Power
Shoshone
Twin Falls
Washington

Illinois
All counties
except:
Alexander
Bond
Christian
Clay
Clinton
Crawford
Edwards
Effingham
Fayette
Franklin
Gallatin
Hamilton
Hardin
Jackson
Jasper
Jefferson
Johnson
Lawrence
Macoupin
Madison
Marion
Massac
Monroe
Montgomery
Perry
Pope
Pulaski
Randolph
Richland
Saline
Shelby
St. Clair
Union
Wabash
Washington
Wayne
White
Williamson

Indiana
All counties
except:
Clark
Crawford
Daviess
Dearborn
Dubois
Floyd
Gibson

Greene
Harrison
Jackson
Jefferson
Jennings
Knox
Lawrence
Martin
Monroe
Ohio
Orange
Perry
Pike
Posey
Ripley
Scott
Spencer
Sullivan
Switzerland
Vanderburgh
Warrick
Washington

Iowa
All counties
except:
Allamakee
Black Hawk
Bremer
Buchanan
Buena Vista
Butler
Calhoun
Cerro Gordo
Cherokee
Chickasaw
Clay
Clayton
Delaware
Dickinson
Emmet
Fayette
Floyd
Franklin
Grundy
Hamilton
Hancock
Hardin
Howard
Humboldt
Ida
Kossuth
Lyon
Mitchell
O'Brien
Osceola
Palo Alto
Plymouth
Pocahontas
Sac
Sioux
Webster
Winnebago
Winneshiek
Worth
Wright

Kansas
Cheyenne
Cloud
Decatur
Ellis
Gove
Graham
Greeley
Hamilton
Jewell
Lane
Logan
Mitchell
Ness
Norton
Osborne
Phillips
Rawlins
Republic
Rooks
Scott
Sheridan
Sherman
Smith
Thomas
Trego
Wallace
Wichita

Maryland
Garrett

Massachusetts
All counties

Michigan
Allegan
Barry
Bay
Berrien
Branch
Calhoun
Cass
Clinton
Eaton
Genesee
Gratiot
Hillsdale
Ingham
Ionia
Jackson
Kalamazoo
Kent
Lapeer
Lenawee
Livingston
Macomb
Midland
Monroe
Montcalm
Muskegon
Oakland
Ottawa
Saginaw

Shiawassee
St. Clair
St. Joseph
Tuscola
Van Buren
Washtenaw
Wayne

Missouri
Adair
Andrew
Atchison
Buchanan
Caldwell
Chariton
Clark
Clinton
Daviess
DeKalb
Gentry
Grundy
Harrison
Holt
Knox
Lewis
Linn
Livingston
Macon
Marion
Mercer
Nodaway
Pike
Putnam
Ralls
Schuyler
Scotland
Shelby
Sullivan
Worth

Nebraska
All counties

Nevada
All counties
except:
Clark

New Hampshire
Cheshire
Hillsborough
Rockingham
Strafford

New Jersey
Bergen
Hunterdon
Mercer
Morris
Passaic
Somerset
Sussex
Warren

New Mexico
Catron
Colfax
Harding
Los Alamos
McKinley
Mora
Rio Arriba
Sandoval
San Juan
San Miguel
Santa Fe
Taos
Torrance

New York
Albany
Cayuga
Chautauqua
Chemung
Columbia
Cortland
Dutchess
Erie
Genesee
Greene
Livingston
Monroe
Niagara
Onondaga
Ontario
Orange
Orleans
Oswego
Putnam
Rensselaer
Rockland
Saratoga
Schenectady
Seneca
Tioga
Washington
Wayne
Yates

North Carolina
Alleghany
Ashe
Avery
Mitchell
Watauga
Yancey

Ohio
All counties
except:
Adams
Brown
Clermont
Gallia
Hamilton
Lawrence
Pike

Scioto
Washington

Oregon
Baker
Crook
Deschutes
Gilliam
Grant
Harney
Hood River
Jefferson
Klamath
Lake
Malheur
Morrow
Sherman
Umatilla
Union
Wallowa
Wasco
Wheeler

Pennsylvania
All counties
except:
Bucks
Cameron
Chester
Clearfield
Delaware
Elk
McKean
Montgomery
Philadelphia
Potter
Susquehanna
Tioga
Wayne
York

Rhode Island
All counties

South Dakota
Bennett
Bon Homme
Charles Mix
Clay
Douglas
Gregory
Hutchinson
Jackson
Mellette
Todd
Tripp
Union
Yankton

Utah
All counties
except:
Box Elder
Cache

Carbon
Daggett
Duchesne
Morgan
Rich
Summit
Uintah
Wasatch
Washington

Washington
Adams
Asotin
Benton
Chelan
Columbia
Douglas
Franklin
Garfield
Grant
Kittitas
Klickitat
Lincoln
Skamania
Spokane
Walla Walla
Whitman
Yakima

Wyoming
Goshen
Platte

West Virginia
Barbour
Brooke
Doddridge
Fayette
Grant
Greenbrier
Hampshire
Hancock
Hardy
Harrison
Lewis
Marion
Marshall
Mineral
Monongalia
Nicholas
Ohio
Pendleton
Pocahontas
Preston
Raleigh
Randolph
Summers
Taylor
Tucker
Upshur
Webster
Wetzel

Climate Zone 5 Recommendation Table for Small Retail Buildings

Item		Component	Recommendation (Minimum or Maximum)	How-To Tips in Chapter 5	✓
Envelope	Roof	Insulation entirely above deck	R-20 c.i.	EN1-2, 17, 20-21	
		Metal building	R-13 + R-19	EN1, 3, 17, 20-21	
		Attic and other	R-38	EN4, 17-18, 20-21	
		Single rafter	R-38 + R-5 c.i.	EN5, 17, 20-21	
		Solar reflectance index (SRI)	No recommendation	EN1	
	Walls	Mass (HC > 7 Btu/ft^2)	R-13.3 c.i.	EN6, 17, 20-21	
		Metal building	R-13 + R-13	EN7, 17, 20-21	
		Steel framed	R-13 + R-7.5 c.i.	EN8, 17, 20-21	
		Wood framed and other	R-13 + R-7.5 c.i.	EN9, 17, 20-21	
		Below-grade walls	R-7.5 c.i.	EN10, 17, 20-21	
	Floors	Mass	R-12.5 c.i.	EN11, 17, 20-21	
		Steel framed	R-30	EN12, 17, 20-21	
		Wood framed and other	R-30	EN12, 17, 20-21	
	Slabs	Unheated	R-10 for 24 in.	EN13, 17, 19-21	
		Heated	R-10 full slab	EN14, 17, 19-21	
	Doors - Opaque	Swinging	U-0.50	EN15, 20-21	
		Non-swinging	U-0.50	EN16, 20-21	
	Vertical Glazing Including Doors	Area (percent of gross wall)	40%	EN22-23, 27, 28, 29	
		Thermal transmittance	U-0.38	EN22, 25	
		Solar heat gain coefficient (SHGC)	N, S, E, W - 0.41; N only—0.41	EN22	
		Exterior sun control (S, E, W only)	Projection factor > 0.5	EN26, DL3	
	Skylights	Area (percent of gross roof)	3%	EN24	
		Thermal transmittance	U-0.69		
		Solar heat gain coefficient (SHGC)	0.36	DL3-10	
Lighting	Interior Lighting	Lighting power density (LPD)	1.3 W/ft^2	EL1, 3, 4, 14, 16-25	
		Linear fluorescent with high-performance electronic ballast	91 mean lm/W	EL7, 8	
		All other sources	50 mean lm/W	EL9,10	
		Dimming controls for daylight harvesting under skylights	Dim fixtures within 10 ft of skylight edge	DL1-9, EL15	
		Occupancy controls	Auto-off all non-sales rooms	EL13	
		Interior room surface reflectances in locations with daylighting	80%+ on ceilings, 70%+ on walls	DL2, EL5	
	Additional Interior Lighting for Sales Floor	Additional LPD for adjustable lighting equipment that is specifically designed and directed to highlight merchandise and is automatically controlled separately from the general lighting	0.4 W/ft^2 (spaces not listed below) 0.6 W/ft^2 (sporting goods, small electronics) 0.9 W/ft^2 (furniture, clothing, cosmetics, and artwork) 1.5 W/ft^2 (jewelry, crystal, china)	EL1, 2, 10, 11, 14, 20, 21	
		Sources	Halogen IR or CMH	EL10-12	
	Exterior Lighting	Façade and externally illuminated signage lighting	0.2 W/ft^2	EX2, 3, EL5, EL7-12	
HVAC	HVAC	Air conditioner (0-65 kBtuh)	13.0 SEER	HV1-4, 6, 12, 16-17, 20	
		Air conditioner (>65-135 kBtuh)	11.0 EER/11.4 IPLV	HV1-4, 6, 12, 16-17, 20	
		Air conditioner (>135-240 kBtuh)	10.8 EER/11.2 IPLV	HV1-4, 6, 12, 16-17, 20	
		Air conditioner (>240 kBtuh)	10.0 EER/10.4 IPLV	HV1-4, 6, 12, 16-17, 20	
		Gas furnace (0-225 kBtuh - SP)	80% AFUE or E_t	HV1-2, 6, 16, 20	
		Gas furnace (0-225 kBtuh - Split)	90% AFUE or E_t	HV1-2, 6, 16, 20	
		Gas furnace (>225 kBtuh)	80% E_c	HV1-2, 6, 16, 20	
		Heat pump (0-65 kBtuh)	13.0 SEER/7.7 HSPF	HV1-4, 6, 12, 16-17, 20	
		Heat pump (>65-135 kBtuh)	10.6 EER/11.0 IPLV/3.2 COP	HV1-4, 6, 12, 16-17, 20	
		Heat pump (>135 kBtuh)	10.1 EER/11.0 IPLV/3.1 COP	HV1-4, 6, 12, 16-17, 20	
	Economizer	Air conditioners & heat pumps- SP	Cooling capacity > 54 kBtuh	HV23	
	Ventilation	Outdoor air damper	Motorized control	HV7-8	
		Demand control	CO_2 sensors	HV7, 22	
	Ducts	Friction rate	0.08 in. w.c./100 ft	HV9, 18	
		Sealing	Seal class B	HV11	
		Location	Interior only	HV9	
		Insulation level	R-6	HV10	
SWH	Service Water Heating	Gas storage (> 75 kBtuh)	90% E_t	WH1-4	
		Gas Instantaneous	0.81 EF or 81% E_t	WH1-4	
		Electric storage (≤ 12 kW and > 20 gal)	EF > 0.99 – 0.0012xVolume	WH1-4	
		Pipe insulation (d < 1½ in./ d ≥ 1½ in.)	1 in./ 1½ in.	WH6	

Note: If the table contains "No recommendation" for a component, the user must meet the more stringent of either Standard 90.1 or the local code requirements in order to reach the 30% savings target.

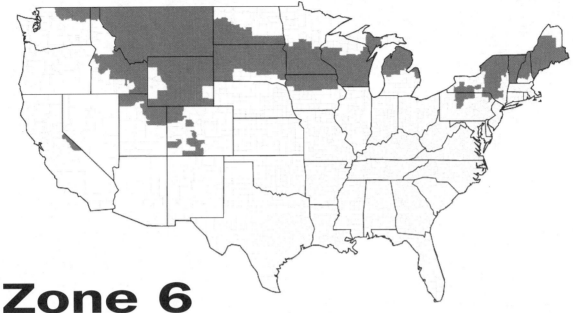

Zone 6

California

Alpine
Mono

Colorado

Alamosa
Archuleta
Chaffee
Conejos
Costilla
Custer
Dolores
Eagle
Moffat
Ouray
Rio Blanco
Saguache
San Miguel

Idaho

Adams
Bannock
Bear Lake
Bingham
Blaine
Boise
Bonner
Bonneville
Boundary
Butte
Camas
Caribou
Clark
Custer
Franklin
Fremont
Jefferson
Lemhi
Madison
Oneida
Teton
Valley

Iowa

Allamakee
Black Hawk

Bremer
Buchanan
Buena Vista
Butler
Calhoun
Cerro Gordo
Cherokee
Chickasaw
Clay
Clayton
Delaware
Dickinson
Emmet
Fayette
Floyd
Franklin
Grundy
Hamilton
Hancock
Hardin
Howard
Humboldt
Ida
Kossuth
Lyon
Mitchell
O'Brien
Osceola
Palo Alto
Plymouth
Pocahontas
Sac
Sioux
Webster
Winnebago
Winneshiek
Worth
Wright

Maine

*All counties
except:*
Aroostook

Michigan

Alcona
Alger

Alpena
Antrim
Arenac
Benzie
Charlevoix
Cheboygan
Clare
Crawford
Delta
Dickinson
Emmet
Gladwin
Grand Traverse
Huron
Iosco
Isabella
Kalkaska
Lake
Leelanau
Manistee
Marquette
Mason
Mecosta
Menominee
Missaukee
Montmorency
Newaygo
Oceana
Ogemaw
Osceola
Oscoda
Otsego
Presque Isle
Roscommon
Sanilac
Wexford

Minnesota

Anoka
Benton
Big Stone
Blue Earth
Brown
Carver
Chippewa
Chisago
Cottonwood
Dakota

Dodge
Douglas
Faribault
Fillmore
Freeborn
Goodhue
Hennepin
Houston
Isanti
Jackson
Kandiyohi
Lac qui Parle
Le Sueur
Lincoln
Lyon
Martin
McLeod
Meeker
Morrison
Mower
Murray
Nicollet
Nobles
Olmsted
Pipestone
Pope
Ramsey
Redwood
Renville
Rice
Rock
Scott
Sherburne
Sibley
Stearns
Steele
Stevens
Swift
Todd
Traverse
Wabasha
Waseca
Washington
Watonwan
Winona
Wright
Yellow Medicine

Montana

All counties

New Hampshire

Belknap
Carroll
Coos
Grafton
Merrimack
Sullivan

New York

Allegany
Broome
Cattaraugus
Chenango
Clinton
Delaware
Essex
Franklin
Fulton
Hamilton
Herkimer
Jefferson
Lewis
Madison
Montgomery
Oneida
Otsego
Schoharie
Schuyler
Steuben
St. Lawrence
Sullivan
Tompkins
Ulster
Warren
Wyoming

North Dakota

Adams
Billings
Bowman
Burleigh
Dickey
Dunn

Emmons
Golden Valley
Grant
Hettinger
LaMoure
Logan
McIntosh
McKenzie
Mercer
Morton
Oliver
Ransom
Richland
Sargent
Sioux
Slope
Stark

Pennsylvania

Cameron
Clearfield
Elk
McKean
Potter
Susquehanna
Tioga
Wayne

South Dakota

*All counties
except:*
Bennett
Bon Homme
Charles Mix
Clay
Douglas
Gregory
Hutchinson
Jackson
Mellette
Todd
Tripp
Union
Yankton

Utah

Box Elder
Cache

Carbon
Daggett
Duchesne
Morgan
Rich
Summit
Uintah
Wasatch

Vermont

All counties

Washington

Ferry
Okanogan
Pend Oreille
Stevens

Wisconsin

*All counties
except:*
Ashland
Bayfield
Burnett
Douglas
Florence
Forest
Iron
Langlade
Lincoln
Oneida
Price
Sawyer
Taylor
Vilas
Washburn

Wyoming

*All counties
except:*
Goshen
Platte
Lincoln
Sublette
Teton

Climate Zone 6 Recommendation Table for Small Retail Buildings

Item		Component	Recommendation (Minimum or Maximum)	How-To Tips in Chapter 5	✓
Envelope	Roof	Insulation entirely above deck	R-20 c.i.	EN1-2, 17, 20-21	
		Metal building	R-13 + R-19	EN1, 3, 17, 20-21	
		Attic and other	R-38	EN4, 17-18, 20-21	
		Single rafter	R-38 + R-5 c.i.	EN5, 17, 20-21	
		Solar reflectance index (SRI)	No recommendation	EN1	
	Walls	Mass (HC > 7 Btu/ft^2)	R-13.3 c.i.	EN6, 17, 20-21	
		Metal building	R-13 + R-13	EN7, 17, 20-21	
		Steel framed	R-13 + R-7.5 c.i.	EN8, 17, 20-21	
		Wood framed and other	R-13 + R-7.5 c.i.	EN9, 17, 20-21	
		Below-grade walls	R-7.5 c.i.	EN10, 17, 20-21	
	Floors	Mass	R-12.5 c.i.	EN11, 17, 20-21	
		Steel framed	R-30	EN12, 17, 20-21	
		Wood framed and other	R-30	EN12, 17, 20-21	
	Slabs	Unheated	R-10 for 24 in.	EN13, 17, 19-21	
		Heated	R-10 full slab	EN14, 17, 19-21	
	Doors – Opaque	Swinging	U-0.50	EN15, 20-21	
		Non-swinging	U-0.50 .	EN16, 20-21	
	Vertical Glazing Including Doors	Area (percent of gross wall)	40%	EN22-23, 27, 28, 29	
		Thermal transmittance	U-0.38	EN22, 25	
		Solar heat gain coefficient (SHGC)	N, S, E, W - 0.41; N only—0.41	EN22	
		Exterior sun control (S, E, W only)	Projection factor > 0.5	EN26, DL3	
	Skylights	Area (percent of gross roof)	3%	EN24	
		Thermal transmittance	U-069		
		Solar heat gain coefficient (SHGC)	0.46	DL3-10	
Lighting	Interior Lighting	Lighting power density (LPD)	1.3 W/ft^2	EL1, 3, 4, 14, 16-25	
		Linear fluorescent with high-performance electronic ballast	91 mean lm/W	EL7, 8	
		All other sources	50 mean lm/W	EL9,10	
		Dimming controls for daylight harvesting under skylights	Dim fixtures within 10 ft of skylight edge	DL1-9, EL15	
		Occupancy controls	Auto-off all non-sales rooms	EL13	
		Interior room surface reflectances in locations with daylighting	80%+ on ceilings, 70%+ on walls	DL2, EL5	
	Additional Interior Lighting for Sales Floor	Additional LPD for adjustable lighting equipment that is specifically designed and directed to highlight merchandise and is automatically controlled separately from the general lighting	0.4 W/ft^2 (spaces not listed below) 0.6 W/ft^2 (sporting goods, small electronics) 0.9 W/ft^2 (furniture, clothing, cosmetics, and artwork) 1.5 W/ft^2 (jewelry, crystal, china)	EL1, 2, 10, 11, 14, 20, 21	
		Sources	Halogen IR or CMH	EL10-12	
	Exterior Lighting	Façade and externally illuminated signage lighting	0.2 W/ft^2	EX2, 3, EL5, EL7-12	
HVAC	HVAC	Air conditioner (0-65 kBtuh)	13.0 SEER	HV1-4, 6, 12, 16-17, 20	
		Air conditioner (>65-135 kBtuh)	No recommendation	HV1-4, 6, 12, 16-17, 20	
		Air conditioner (>135-240 kBtuh)	No recommendation	HV1-4, 6, 12, 16-17, 20	
		Air conditioner (>240 kBtuh)	No recommendation	HV1-4, 6, 12, 16-17, 20	
		Gas furnace (0-225 kBtuh - SP)	80% AFUE or E_t	HV1-2, 6, 16, 20	
		Gas furnace (0-225 kBtuh - Split)	90% AFUE or E_t	HV1-2, 6, 16, 20	
		Gas furnace (>225 kBtuh)	80% E_c	HV1-2, 6, 16, 20	
		Heat pump (0-65 kBtuh)	13.0 SEER/7.7 HSPF	HV1-4, 6, 12, 16-17, 20	
		Heat pump (>65-135 kBtuh)	No recommendation	HV1-4, 6, 12, 16-17, 20	
		Heat pump (>135 kBtuh)	No recommendation	HV1-4, 6, 12, 16-17, 20	
	Economizer Ventilation	Air conditioners & heat pumps- SP	Cooling capacity > 54 kBtuh	HV23	
		Outdoor air damper	Motorized control	HV7-8	
		Demand control	CO_2 sensors	HV7, 22	
	Ducts	Friction rate	0.08 in. w.c./100 ft	HV9, 18	
		Sealing	Seal class B	HV11	
		Location	Interior only	HV9	
		Insulation level	R-6	HV10	
SWH	Service Water Heating	Gas storage (> 75 kBtuh)	90% E_t	WH1-4	
		Gas Instantaneous	0.81 EF or 81% E_t	WH1-4	
		Electric storage (≤ 12 kW and > 20 gal)	EF > 0.99 – 0.0012xVolume	WH1-4	
		Pipe insulation (d < 1½ in./ d ≥ 1½ in.)	1 in./ 1½ in.	WH6	

Note: If the table contains "No recommendation" for a component, the user must meet the more stringent of either Standard 90.1 or the local code requirements in order to reach the 30% savings target.

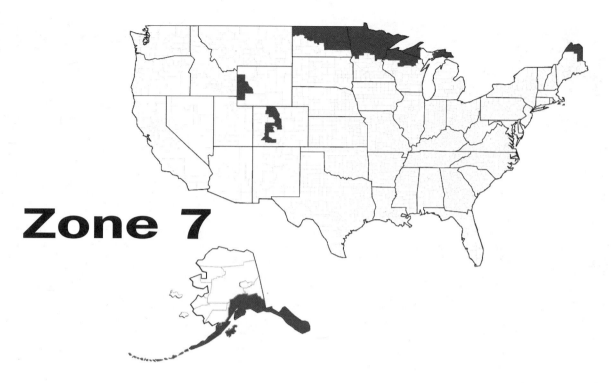

Zone 7

Alaska

Aleutians East
Aleutians West (CA)
Anchorage
Angoon (CA)
Bristol Bay
Denali
Haines
Juneau
Kenai Peninsula
Ketchikan (CA)
Ketchikan Gateway
Kodiak Island
Lake and Peninsula
Matanuska-Susitna
Prince of Wales-Outer
Sitka
Skagway-Hoonah-
Valdez-Cordova (CA)
Wrangell-Petersburg (CA)
Yakutat

Colorado

Clear Creek
Grand
Gunnison
Hinsdale
Jackson
Lake
Mineral
Park
Pitkin
Rio Grande
Routt

San Juan
Summit

Maine

Aroostook

Michigan

Baraga
Chippewa
Gogebic
Houghton
Iron
Keweenaw
Luce
Mackinac
Ontonagon
Schoolcraft

Minnesota

Aitkin
Becker
Beltrami
Carlton
Cass
Clay
Clearwater
Cook
Crow Wing
Grant
Hubbard
Itasca
Kanabec
Kittson
Koochiching

Lake
Lake of the Woods
Mahnomen
Marshall
Mille Lacs
Norman
Otter Tail
Pennington
Pine
Polk
Red Lake
Roseau
St. Louis
Wadena
Wilkin

North Dakota

Barnes
Benson
Bottineau
Burke
Cass
Cavalier
Divide
Eddy
Foster
Grand Forks
Griggs
Kidder
McHenry
McLean
Mountrail
Nelson
Pembina
Pierce
Ramsey

Renville
Rolette
Sheridan
Steele
Stutsman
Towner
Traill
Walsh
Ward
Wells
Williams

Wisconsin

Ashland
Bayfield
Burnett
Douglas
Florence
Forest
Iron
Langlade
Lincoln
Oneida
Price
Sawyer
Taylor
Vilas
Washburn

Wyoming

Lincoln
Sublette
Teton

Climate Zone 7 Recommendation Table for Small Retail Buildings

Item	Component	Recommendation (Minimum or Maximum)	How-To Tips in Chapter 5	✓
Roof	Insulation entirely above deck	R-25 c.i.	EN1-2, 17, 20-21	
	Metal building	R-16 + R-19	EN1, 3, 17, 20-21	
	Attic and other	R-60	EN4, 17-18, 20-21	
	Single rafter	R-38 + R-10 c.i.	EN5, 17, 20-21	
	Solar reflectance index (SRI)	No recommendation	EN1	
Walls	Mass (HC > 7 Btu/ft^2)	R-15.2 c.i.	EN6, 17, 20-21	
	Metal building	R-13 + R-13	EN7, 17, 20-21	
	Steel framed	R-13 + R-7.5 c.i.	EN8, 17, 20-21	
	Wood framed and other	R-13 + R-7.5 c.i.	EN9, 17, 20-21	
	Below-grade walls	R-7.5 c.i.	EN10, 17, 20-21	
Floors	Mass	R-14.6 c.i.	EN11, 17, 20-21	
	Steel framed	R-38	EN12, 17, 20-21	
	Wood framed and other	R-30	EN12, 17, 20-21	
Slabs	Unheated	R-15 for 24 in.	EN13, 17, 19-21	
	Heated	R-15 full slab	EN14, 17, 19-21	
Doors – Opaque	Swinging	U-0.50	EN15, 20-21	
	Non-swinging	U-0.50	EN16, 20-21	
Vertical Glazing Including Doors	Area (percent of gross wall)	40%	EN22-23, 27, 28, 30	
	Thermal transmittance	U-0.38	EN22, 25	
	Solar heat gain coefficient (SHGC)	N, S, E, W - 0.41; N only—0.41	EN22	
	Exterior sun control (S, E, W only)	Projection factor > 0.5	EN26, DL3	
Skylights	Area (percent of gross roof)	3%	EN24	
	Thermal transmittance	U-0.69		
	Solar heat gain coefficient (SHGC)	0.64	DL3-10	
Interior Lighting	Lighting power density (LPD)	1.3 W/ft^2	EL1, 3, 4, 14, 16-25	
	Linear fluorescent with high-performance electronic ballast	91 mean lm/W	EL7, 8	
	All other sources	50 mean lm/W	EL9,10	
	Dimming controls for daylight harvesting under skylights	Dim fixtures within 10 ft of skylight edge	DL1-9, EL15	
	Occupancy controls	Auto-off all non-sales rooms	EL13	
	Interior room surface reflectances in locations with daylighting	80%+ on ceilings, 70%+ on walls	DL2, EL5	
Additional Interior Lighting for Sales Floor	Additional LPD for adjustable lighting equipment that is specifically designed and directed to highlight merchandise and is automatically controlled separately from the general lighting	0.4 W/ft^2 (spaces not listed below) 0.6 W/ft^2 (sporting goods, small electronics) 0.9 W/ft^2 (furniture, clothing, cosmetics, and artwork) 1.5 W/ft^2 (jewelry, crystal, china)	EL1, 2, 10, 11, 14, 20, 21	
	Sources	Halogen IR or CMH	EL10-12	
Exterior Lighting	Façade and externally illuminated signage lighting	0.2 W/ft^2	EX2, 3, EL5, EL7-12	
HVAC	Air conditioner (0-65 kBtuh)	13.0 SEER	HV1-4, 6, 12, 16-17, 20	
	Air conditioner (>65-135 kBtuh)	No recommendation	HV1-4, 6, 12, 16-17, 20	
	Air conditioner (>135-240 kBtuh)	No recommendation	HV1-4, 6, 12, 16-17, 20	
	Air conditioner (>240 kBtuh)	No recommendation	HV1-4, 6, 12, 16-17, 20	
	Gas furnace (0-225 kBtuh - SP)	80% AFUE or E$_t$	HV1-2, 6, 16, 20	
	Gas furnace (0-225 kBtuh - Split)	90% AFUE or E$_t$	HV1-2, 6, 16, 20	
	Gas furnace (>225 kBtuh)	80% E$_c$	HV1-2, 6, 16, 20	
	Heat pump (0-65 kBtuh)	13.0 SEER/7.7 HSPF	HV1-4, 6, 12, 16-17, 20	
	Heat pump (>65-135 kBtuh)	No recommendation	HV1-4, 6, 12, 16-17, 20	
	Heat pump (>135 kBtuh)	No recommendation	HV1-4, 6, 12, 16-17, 20	
Economizer	Air conditioners & heat pumps- SP	No recommendation	HV23	
Ventilation	Outdoor air damper	Motorized control	HV7-8	
	Demand control	CO_2 sensors	HV7, 22	
Ducts	Friction rate	0.08 in. w.c./100 ft	HV9, 18	
	Sealing	Seal class B	HV11	
	Location	Interior only	HV9	
	Insulation level	R-6	HV10	
Service Water Heating	Gas storage (> 75 kBtuh)	90% E$_t$	WH1-4	
	Gas Instantaneous	0.81 EF or 81% E$_t$	WH1-4	
	Electric storage (≤ 12 kW and > 20 gal)	EF > 0.99 – 0.0012xVolume	WH1-4	
	Pipe insulation (d < 1½ in./ d ≥ 1½ in.)	1 in./ 1½ in.	WH6	

Note: If the table contains "No recommendation" for a component, the user must meet the more stringent of either Standard 90.1 or the local code requirements in order to reach the 30% savings target.

Zone 8

Alaska

Bethel (CA)
Dillingham (CA)
Fairbanks North Star
Nome (CA)
North Slope
Northwest Arctic
Southeast Fairbanks (CA)
Wade Hampton (CA)
Yukon-Koyukuk (CA)

Climate Zone 8 Recommendation Table for Small Retail Buildings

	Item	Component	Recommendation (Minimum or Maximum)	How-To Tips in Chapter 5	✓
Envelope	Roof	Insulation entirely above deck	R-25 c.i.	EN1-2, 17, 20-21	
		Metal building	R-16 + R-19	EN1, 3, 17, 20-21	
		Attic and other	R-60	EN4, 17-18, 20-21	
		Single rafter	R-38 + R-10 c.i.	EN5, 17, 20-21	
		Solar reflectance index (SRI)	No recommendation	EN1	
	Walls	Mass (HC > 7 Btu/ft^2)	R-15.2 c.i.	EN6, 17, 20-21	
		Metal building	R-13 + R-13	EN7, 17, 20-21	
		Steel framed	R-13 + R-10 c.i.	EN8, 17, 20-21	
		Wood framed and other	R-13 + R-7.5 c.i.	EN9, 17, 20-21	
		Below-grade walls	R-7.5 c.i.	EN10, 17, 20-21	
	Floors	Mass	R-14.6 c.i.	EN11, 17, 20-21	
		Steel framed	R-38	EN12, 17, 20-21	
		Wood framed and other	R-30	EN12, 17, 20-21	
	Slabs	Unheated	R-15 for 24 in.	EN13, 17, 19-21	
		Heated	R-15 full slab	EN14, 17, 19-21	
	Doors-Opaque	Swinging	U-0.50	EN15, 20-21	
		Non-swinging	U-0.50	EN16, 20-21	
	Vertical Glazing Including Doors	Area (percent of gross wall)	40%	EN22-23, 27, 28, 30	
		Thermal transmittance	U-0.38	EN22, 25	
		Solar heat gain coefficient (SHGC)	N, S, E, W - 0.41; N only—0.41	EN22	
		Exterior sun control (S, E, W only)	Projection factor > 0.5	EN26, DL3	
	Skylights	Area (percent of gross roof)	3%	EN24	
		Thermal transmittance	U-0.58		
		Solar heat gain coefficient (SHGC)	0.64	DL3-10	
Lighting	Interior Lighting	Lighting power density (LPD)	1.3 W/ft^2	EL1, 3, 4, 14, 16-25	
		Linear fluorescent with high-performance electronic ballast	91 mean lm/W	EL7, 8	
		All other sources	50 mean lm/W	EL9,10	
		Dimming controls for daylight harvesting under skylights	Dim fixtures within 10 ft of skylight edge	DL1-9, EL15	
		Occupancy controls	Auto-off all non-sales rooms	EL13	
		Interior room surface reflectances in locations with daylighting	80%+ on ceilings, 70%+ on walls	DL2, EL5	
	Additional Interior Lighting for Sales Floor	Additional LPD for adjustable lighting equipment that is specifically designed and directed to highlight merchandise and is automatically controlled separately from the general lighting	0.4 W/ft^2 (spaces not listed below) 0.6 W/ft^2 (sporting goods, small electronics) 0.9 W/ft^2 (furniture, clothing, cosmetics, and artwork) 1.5 W/ft^2 (jewelry, crystal, china)	EL1, 2, 10, 11, 14, 20, 21	
		Sources	Halogen IR or CMH	EL10 – 12	
	Exterior Lighting	Façade and externally illuminated signage lighting	0.2 W/ft^2	EX2, 3, EL5, EL7-12	
HVAC	HVAC	Air conditioner (0-65 kBtuh)	13.0 SEER	HV1-4, 6, 12, 16-17, 20	
		Air conditioner (>65-135 kBtuh)	No recommendation	HV1-4, 6, 12, 16-17, 20	
		Air conditioner (>135-240 kBtuh)	No recommendation	HV1-4, 6, 12, 16-17, 20	
		Air conditioner (>240 kBtuh)	No recommendation	HV1-4, 6, 12, 16-17, 20	
		Gas furnace (0-225 kBtuh - SP)	80% AFUE or E_t	HV1-2, 6, 16, 20	
		Gas furnace (0-225 kBtuh - Split)	90% AFUE or E_t	HV1-2, 6, 16, 20	
		Gas furnace (>225 kBtuh)	80% E_c	HV1-2, 6, 16, 20	
		Heat pump (0-65 kBtuh)	13.0 SEER/7.7 HSPF	HV1-4, 6, 12, 16-17, 20	
		Heat pump (>65-135 kBtuh)	No recommendation	HV1-4, 6, 12, 16-17, 20	
		Heat pump (>135 kBtuh)	No recommendation	HV1-4, 6, 12, 16-17, 20	
	Economizer	Air conditioners & heat pumps- SP	No recommendation	HV23	
	Ventilation	Outdoor air damper	Motorized control	HV7-8	
		Demand control	CO_2 sensors	HV7, 22	
	Ducts	Friction rate	0.08 in. w.c./100 ft	HV9, 18	
		Sealing	Seal class B	HV11	
		Location	Interior only	HV9	
		Insulation level	R-8	HV10	
SWH	Service Water Heating	Gas storage (> 75 kBtuh)	90% E_t	WH1-4	
		Gas instantaneous	0.81 EF or 81% E_t	WH1-4	
		Electric storage (≤ 12 kW and > 20 gal)	EF > 0.99 – 0.0012xVolume	WH1-4	
		Pipe insulation (d < 1½ in./ d ≥ 1½ in.)	1 in./ 1½ in.	WH6	

Note: If the table contains "No recommendation" for a component, the user must meet the more stringent of either Standard 90.1 or the local code requirements in order to reach the 30% savings target.

Technology Examples and Case Studies 4

Chapter 4 presents technology examples and case studies of buildings that illustrate the principles of the recommendations presented in this Guide. The technology examples represent a specific portion of the building that would contribute toward meeting the energy target—such as lighting, equipment efficiencies, or envelope measures. The case studies are buildings that meet the energy target.

CLIMATE ZONE 2—HAPPY FEET PLUS

CLEARWATER, FLORIDA

Happy Feet Plus is an approximately 6,000 ft² retail shoe store located in Clearwater, Florida (climate zone 2). The project was constructed in a design-build format,

Figure 4-1. Happy Feet Plus exterior.

(a)   (b)

Figure 4-2. Happy Feet Plus (a) interior and (b) sales floor. Note the entry vestibule and ceiling fans and the ceiling-mounted fluorescent fixtures for merchandise display lighting.

with the owner providing the conceptual design and layout of the store. Approximately 50% of the space is owner-occupied, and the rest is leased.

Energy efficiency for the store relies upon a very high-performance building envelope and a high-performance HVAC system. The energy code requires no insulation for massive walls in this area, while the insulating concrete form system provides R-40. The acoustical, insulated roof panel provides an R-30 roof, while the code requirement is R-19. The HVAC system utilizes a high-efficiency split heat pump and an enthalpy wheel-type energy recovery ventilator (ERV). In addition to reducing energy costs, the ERV reduces the peak cooling demand of the facility to only 5 tons of refrigeration.

The lighting system for the showroom utilizes T-8 fluorescent lamps, electronic ballasts, and surface-mounted high-efficiency light fixtures to provide an average of 75 footcandles at table level with only 1.0 W per square foot. The photovoltaic (PV) system is expected to provide almost 50% of the building's total energy needs. These measures provide energy savings for the building of well over 60% compared with a minimally code-compliant building. Actual savings may be even greater. Energy bills for the first year of operation of the owner-occupied space were only about $1/\text{ft}^2$.

In addition to energy efficiency measures, the project incorporates a rain catchment system to provide all water for irrigation and sewage conveyance, utilizing a 5000 gal underground cistern supplied by the roof drains. The parking area is paved with a permeable paving to reduce stormwater runoff. The project received LEED Gold Certification.

The following how-to tips were implemented in this project: EN1, EN4, EN6, EN15, EN20, EN23, DL3, EL1, EL2, EL6, EL7, EL14, EL18, EL19, EL27, HV1, HV2, HV3, HV5, HV6, HV7, HV9, HV14, and HV15.

HAPPY FEET PLUS	
Processes for Achieving Energy Savings	**Description of Project Elements**
Envelope	
Opaque Envelope Components	Walls—Insulating concrete forms with cast-in-place concrete—up to R-40. Roof—Acoustical, insulated panels R-30. Doors—U—0.34.
Vertical Glazing (Envelope)	Window U-factor—0.62. Window SHGC—0.6. Window-to-wall ratio—11%.
Lighting	
Electric Lighting Design	Lighting power allowance—1.08 W/ft^2.
HVAC	
Equipment	Split system heat pump—14.75 SEER. Energy recovery ventilator—enthalpy wheel.
Ventilation	Ventilation air controllable by ERV fan controls.
Controls	Programmable thermostats.
Service Water Heating	
SWH	Electric.
Additional Savings	
Other	4.0 kW grid-coupled PV array.

CLIMATE ZONE 3—INTERFACE SHOWROOM

ATLANTA, GEORGIA

The Interface Showroom, opened to the public in 2004, is located within an urban renewal district in Midtown Atlanta, which is in climate zone 3, and includes commercial and retail space for display of carpet. It was designed and built with the goal of being designated a Platinum Project under the USGBC's LEED-CI pilot program; it did achieve that goal. As the Interface Showroom was a tenant space in a new building, upgrades to the HVAC were not possible; therefore, most energy-related sustainable initiatives focused on Cx, optimizing lighting, equipment, and appliances, measurement and verification, and green power.

Photograph courtesy of Brian Gassel/TVS

Figure 4-3. Interface Showroom building exterior showing first floor retail space of existing base building.

Photograph courtesy of Gensler

(a)

Photograph courtesy of Brian Gassel/TVS

(b)

Figure 4-4. (a) Main entrance to retail space, which is flexible to facilitate displaying carpet under varying interior lighting conditions. (b) Area for product presentations or meetings, shown with curtains pulled to enclose the space.

The lighting in this space is truly unique for a retail showroom application. The space utilizes predominately fluorescent and metal halide lighting. These light sources significantly reduce the watts per square foot energy usage for lighting. In addition, the lighting is on an astronomical timer that significantly drops lighting levels after hours. A PV cell monitors the amount of daylight coming in the windows and adjusts electrical lighting accordingly. Occupancy sensors for lighting are located in all spaces that are not regularly occupied. There are a variety of lighting levels in the office space, including natural daylight, ceiling fluorescents, and desk task lighting, to allow designers previewing projects for clients to simulate a number of real-life lighting scenarios.

The lighting system design resulted in a 41% improvement over the applicable energy standards (ASHRAE Standard 90.1-1999). This translates to a 1.15 W/ft^2 actual lighting power density versus a 1.9 W/ft^2 allowance.

Additional Features

Interface provided carbon dioxide monitoring for the entire first floor of the existing base building. The entire west-facing curtain wall, shaded with exterior canopies, gives access to pedestrian-level daylight and views. In addition, the office area contains a low wall, a movable partition system that does not obscure views.

The following how-to tips were implemented in this project: QA1, QA3, QA6, QA8A, QA10, QA12, QA16, EN26, DL1, DL2, DL6, DL7, DL9, EL1, EL2, EL3, EL5, EL9, EL10, EL13, EL14, EL18, EL19, EL21, EL23, EL25, HV7, HV22, WH1, WH2, WH3, WH5, and WH6.

INTERFACE SHOWROOM	
Processes for Achieving Energy Savings	**Description of Project Elements**
Lighting	
Window Design for Daylight	Exterior canopies.
Daylighting	A photocell located along the exterior window wall reads available daylight and, in conjunction with the daylight interface module, adjusts the interior lighting when adequate daylight is available. This helps balance the lighting as well as provide energy savings through daylight harvesting.
Electric Lighting Design	Most lighting fixtures are fluorescent or metal halide to maximize efficiency. Fluorescent lamping is a combination of T-5HO (high output), T-8, and compact fluorescent. The compact fluorescent lamps were used for downlights, as well as for decorative wall sconces and table lamps. Linear fluorescent (T-5HO and T-8) were used for architectural cove lighting, illumination of "super-graphics," and special large-scale architectural "lamp shade" features. With the exception of office work areas and storage rooms, the fluorescent fixtures are all dimmable. Low-wattage CMH lamping was used for accent and display lighting. The lighting design and controls incorporate multiple control zones and various color temperature lamps to simulate conditions of residential (3000 K), commercial/office (3500 K), and natural lighted environments (4100 K).
HVAC	
Controls	CO_2 sensors for ventilation control.
Service Water Heating	
Tenant improvement of new space	15 gal, 2.5 kW electric storage water heater, ENERGY STAR® rated.

CLIMATE ZONE 3—PETCO ENERGY SHOWCASE STORE

LAKE ELSINORE, CALIFORNIA

The PETCO Energy Showcase Store is located in Lake Elsinore, California, in climate zone 3. It consists of a 17,500 ft^2 retail pet supply store located in a strip mall. The building was designed to showcase an extensively daylit big-box retail building.

Energy-saving features include retail floor lighting designed to 1.2 W/ft^2, tubular daylighting devices that provide daylighting in the retail sales area as well as for the

Photograph courtesy Neall Digert

Figure 4-5. Exterior view of PETCO Energy Showcase Store.

Photographs courtesy Neall Digert

(a)   (b)

Figure 4-6. PETCO Energy Showcase Store (a) sales floor and (b) grooming room, both showing daylighting.

point-of-sale and back-of-house areas, a five-zone digital addressable lighting interface controls system to continuously dim lights in response to available daylighting, occupancy sensors for restroom and office lighting, high-efficiency and instantaneous service hot water systems, outdoor air controlled by a CO_2 sensor combined with an economizer cycle, and a reflective cool roof.

Tubular daylighting devices at a density of 200 ft^2 per device provide daylighting to every space in the store except for the aquatic area in the center of the store. Design simulations predict 20% whole-building energy savings from the tubular daylighting devices and daylighting controls. Additional lighting savings are expected from occupancy sensors in the restrooms, pre-sales area, and offices.

Additional Features

The store also has a wireless energy management system for controlling HVAC and lighting systems as well as for monitoring additional temperature and status points. The ease of installing additional wireless sensors allows for aggressive load management by controlling lighting setpoints, cooling setpoints, and other equipment loads during a peak demand event.

The following how-to tips were implemented in this project: QA8A, QA10, QA16, EN26, DL1, DL3, DL4, DL6, DL7, DL8, EL1, EL2, EL3, EL7, EL10, EL13, EL14, EL15, EL16, EL19, EL21, EL24, EL28, HV14, HV21, HV22, HV23, WH1, WH2, WH4, PL4, and EX4.

PETCO ENERGY SHOWCASE STORE	
Processes for Achieving Energy Savings	**Description of Project Elements**
Envelope	
Opaque Envelope Components	High-reflectance cool roof.
Window Design for Thermal Conditions	Storefront windows provide daylighting to point of sale and vestibule.
Window Design for Daylight	Overhangs on storefront windows.
Lighting	
Daylighting	Tubular daylighting devices provide daylighting to 80% of the building; five lighting zones controlled by digitally addressable lighting interface and photocells and occupancy sensors.
Electric Lighting Design	1.2 W/ft^2 T-8s and metal halide lamps, display and aquarium lighting automatically turned off during unoccupied hours.
HVAC	
Equipment	Five packaged rooftop units with outdoor air controlled by CO_2, integrated economizer cycle.
Service Water Heating	
SWH	60 gal, high-efficiency (94%) gas service hot water, instantaneous gas hot water system for restrooms.
Additional Savings	
Other	Peak demand shedding with wireless energy management system.

CLIMATE ZONE 3—REAL GOODS SOLAR LIVING CENTER

HOPLAND, CALIFORNIA

Real Goods Trading Corporation, a distributor of energy conservation and self-sufficiency products, built the Solar Living Center to be a showroom to mirror its retailing ethic. The 5,470 ft^2 center is located in Hopland, California (climate zone 3), and was completed in April 1996. The showroom at the Real Goods Solar Living Center has incorporated various energy-conserving passive control strategies, such as nighttime ventilation, thermal mass, a well-insulated envelope, stack effect natural ventilation, and passive solar heating. The building is arranged in a curved plan that looks like a sundial with a curved courtyard and stepped roofs where clerestory windows capture the varying hourly and seasonal angles of the sun. Daylighting illuminates the facility through the use of clerestory windows, light shelves, trellises, and manually adjustable awnings. Interior walls are painted white for greatest reflectivity. Fluorescent T-8 lighting at 0.6 W/ft^2 is available but rarely needed.

The building's curved back walls are constructed of 23 in. wide straw bales with 3–4 in. of gun-earth on each side to provide an R-value of 65 as well as a significant thermal mass. The roof is constructed with R-50 continuous insulation (c.i.) above-

Photographs courtesy of Jeff Oldham/Real Goods

Figure 4-7. The landscaping in front of Real Goods Solar Living Center.

(a)

(b)

Figure 4-8. (a) The entry elevation of Real Goods Solar Living Center makes use of overhangs, a recycled-redwood trellis, and adjustable awnings. (b) The interior reveals its curving, stepped roof, east-facing clerestory windows, and light shelves that deliver natural light deeper into the interior without overheating or glare.

deck insulation, and the slab perimeter and footings include R-10 c.i. As the building has no mechanical heating or cooling system, natural ventilation, thermal mass, and passive solar gains are used for space conditioning. In the winter months, the low sun penetrates deep into the building to provide warmth and light. In the summer, overhangs and awnings control solar gain. Operable windows with low-e glazing allow natural ventilation and help to reduce heat gain. Backup heating for the coldest winter mornings is supplied by wood stoves. These systems also flush the building with cool night air and then store the coolness in the thermal mass for later use.

All noncritical loads are manually turned off at night. An operators' manual was provided to the Real Goods Solar Living Center staff to teach them how to properly tune the building for optimal comfort and efficiency.

Additional Features

The center's energy-production system, connected to the electric grid, generates 14 kW of PV power and 1 KW of wind power, more than enough to power the site. Extra energy is sold to the local energy company.

The following how-to tips were implemented in this project: QA1, QA3, QA16, EN2, EN6, EN11, EN19, EN24, EN25, EN26, DL1, DL2, DL3, DL7, DL9, EL1, EL5, EL7, EL15, EL19, EL26, PL1, PL2, and EX1.

REAL GOODS SOLAR LIVING CENTER	
Processes for Achieving Energy Savings	**Description of Project Elements**
Envelope	
Opaque Envelope Components	R-50 c.i. above-deck insulation (14 in. cellulose, aluminum radiant barrier, vent air space). R-65 insulated cavity (straw bale). R-10 c.i. under and along entire 3 ft perimeter slab floor and vertical along footings; center of slab is earth coupled; 600 ton mass in 5000 ft^2.
Vertical Glazing (Envelope)	U-factor 0.72, 26% window-wall ratio (passive winter heating).
Window Design for Thermal Conditions	0.68 SHGC, clerestory and low windows opened and closed by staff as part of the natural ventilation system. All glazing faces south except for clerestory, which faces east.
Window Design for Daylight	Overhangs on all glazing, fixed and operable.
Lighting	
Daylighting	East clerestories, white curved ceiling, full height south curtain wall, interior and exterior light shelves. Interior is insulated and operable to cover upper curtain wall. Continuous dimming controls based on interior photocells.
Electric Lighting Design	0.6 W/ft^2 pendant-type, T-8 fluorescent fixtures.
HVAC	
Equipment	No mechanical HVAC, wood stoves for backup heating.
Ventilation	Natural ventilation through staff-operated clerestory windows and open front and back doors.
Service Water Heating	
SWH	35 gal passive solar hot water heater for the restrooms and service sinks.
Additional Savings	
Plug Loads	All noncritical loads manually turned off at night. All wires in conduit and twisted to mitigate EMFs.
Exterior Lighting	CFL, manual control on all night; total of 130 W.
Other	10 kW of PV on trackers plus 4 kW of building-integrated PV, 1 kW demonstration wind turbine, free-standing. Battery backup for eight hours without sun or wind.
Water	Grey water recycling for irrigation.

CLIMATE ZONE 4—NUSTA SPA

WASHINGTON, DC

Nusta Spa is a full-service day spa located in downtown Washington, DC, which is in climate zone 4. This 4,600 ft^2 building retrofit project called for new electrical, mechanical, and fire protection systems. Nusta Spa is the first spa to be accepted into the USGBC's Leadership in Energy and Environmental Design for Commercial Interiors (LEED-CI) pilot program and is LEED-CI Gold certified.

The Nusta Spa space was completely disconnected from the existing base building constant-volume system and equipped with a new 20 ton air-cooled chiller and associated horizontal fan-coil units (FCUs) with electric heat. This increased the available

Figure 4-9. Nusta Spa main entrance.

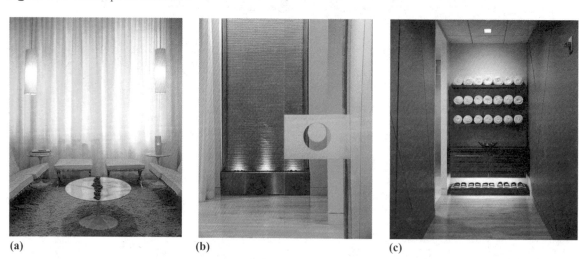

(a)　　　　　　　　　　(b)　　　　　　　　　　(c)

Figure 4-10. (a) Natural lighting with occupancy sensors, (b) a recycled water fountain, and (c) lighting design of 1.3 W/ft^2.

control and allowed individual treatment rooms to maintain environments matching particular clients' preferences. In addition, an ERV enthalpy wheel was used to exchange heat/humidity between the outgoing exhaust stream and the incoming outside airstream and a humidifier was added in the outside air FCU to add humidity to all rooms. ASHRAE 90.1-2001 has a lighting allowance of 1.9 W/ft^2 for retail areas. Using this lighting design achieved 1.3 W/ft^2, roughly 30% lower than the standard requirements. Because of the high hot water demand for the spa, a large hot water storage tank was installed coupled with an 84% efficient noncondensing boiler. This is the highest efficiency that can be achieved without utilizing a condensing boiler.

Throughout the space, eco-friendly strategies are used, including local and regional materials, FSC-certified and reclaimed wood, rapidly renewable materials, low-emitting materials, recycled materials, and energy-efficient lighting and HVAC systems. Exceptional finishes were also installed throughout, including ceramic and glass tiles, fabrics, wood floors, and millwork. An 11 ft tall wall of oak boards milled from reclaimed structural timbers defines the separation between public and private spaces.

The following how-to tips were implemented in this project: QA1, QA3, QA8A, QA16, EL1, EL2, EL3, EL5, HV5, HV7, HV13, HV14, WH1, WH2, and WH3.

NUSTA SPA	
Processes for Achieving Energy Savings	**Description of Project Elements**
Envelope	
This was a building commercial interiors retrofit	No change, existing envelope.
Lighting	
Electric Lighting Design	ASHRAE 90.1-2001 had a lighting allowance of 1.9 W/ft^2 for retail areas; this lighting design achieved 1.3 W/ft^2, roughly 30% lower than the standard requirements.
HVAC	
Equipment	A 20 ton air-cooled chiller and associated horizontal FCUs with electric heat, an enthalpy wheel to exchange heat/humidity between the outgoing exhaust stream and the incoming outside airstream, and a humidifier in the outside air FCU.
Ventilation	Ventilation system configured so that one FCU provided outside air to the entire tenant space; FCU also has variable-speed capability to adjust the outside air delivered to the space based on the average CO_2 reading in the spa, which limited the ventilation air heating and cooling loads on the system to those necessary to keep the occupants comfortable, as opposed to the maximum required per code.
Controls	Each FCU was provided with a digital microprocessor control thermostat that enables individual clients to adjust the temperatures of the spaces to their preferences. The controllers also have four separate time-of-day settings that allow the spa to have a night setback, an occupied mode, and "swing" periods for warming up the spa in the morning and cooling it off at night (see above section for description of CO_2 control).
Service Water Heating	
SWH	250 gal hot water storage tank coupled with an 84% efficient noncondensing boiler.

CLIMATE ZONE 5—ZION NATIONAL PARK VISITOR CENTER AND BOOKSTORE

SPRINGDALE, UTAH

The Visitor Center Complex at Zion National Park in southwestern Utah is located in a dry region of climate zone 5. The 8,800 ft^2 complex contains a retail bookstore, visitor orientation and support areas, and a 2,756 ft^2 restroom facility. It incorporates energy-efficient features such as clerestory daylighting, enhanced envelope, occupancy sensors, solar load control with engineered overhangs, and computerized building controls. Daylighting is provided by south- and east-facing clerestories and north-view glass. The lighting density in the bookstore is 0.9 W/ft^2. The energy management system controls the lights by turning off lamps in response to available daylight. Motion sensors in the offices and support areas turn on the lights when these spaces are occupied. Envelope features include 6 in. steel-stud walls with R-21 expanding blown-in-place foam insulation and exterior 1.5 in. continuous insulating extruded foam and a roof of structural-insulated panels with a continuous R-value of 30.9. Overhangs shade south and east glass from the high sun and shield the facility from unnecessary solar gains during the summer months.

Additional Features

The innovative heating and cooling systems eliminate all ductwork and on-site fuel storage. Passive solar heating and trombe walls augment localized electric heating systems to meet the heating needs. The electric heating systems are controlled to purchase electricity when demand charges will not be incurred. In addition, the energy manage-

Photograph courtesy of DOE/NREL

Figure 4-11. Zion National Park Visitor Center and Bookstore entryway showing the downdraft cooltowers and overhangs.

Photograph courtesy of DOE/NREL

Figure 4-12. Zion Bookstore showing daylighting and 0.9 W/ft^2 of electrical lighting.

ment system delays the use of the electric service hot water system to shift this load to off-peak periods. The cooltowers use a wetted medium at the top of a tower to eliminate the need for conventional air conditioning—cool air naturally falls down the tower and into the building without fans. An uninterruptible power supply (UPS) system is integrated with a 7.2 kW PV system. The UPS enables the bookstore to operate its cash registers, telephones, and security system during the park's frequent power outages.

The integrated design resulted in a building that costs $0.43/ft^2 to operate and consumes 27.0 kBtu/ft^2—a 65% energy savings compared to a similar store built to meet the minimum requirements of ASHRAE Standard 90.1-1999.

The building uses very little energy for heating and cooling when the outdoor temperature is between 60°F and 75°F. The design team set and committed to energy performance goals early in the process, used energy simulation models to predict energy performance and guide design decisions, and continuously monitored, evaluated, and improved performance.

The following how-to tips were implemented in this project: QA1, QA3, QA8, QA9, AQ10, QA16, EN2, EN8, EN13, EN22, EN23, EN26, EN 27, EN 29, DL1, DL2, DL3, DL7, DL8, DL9, DL10, EL5, EL7, EL9, EL10, EL12, EL13, EL14, EL15, EL16, EL19, EL28, HV13, HV14, HV21, PL1, PL4, PL4, EX2, EX4.

ZION NATIONAL PARK VISITOR CENTER AND BOOKSTORE	
Processes for Achieving Energy Savings	**Description of Project Elements**
Envelope	
Opaque Envelope Components	Roof R-30 c.i. SIPs. 6 in. steel-framed walls with R-21 expanding foam cavity insulation with 1-1/2 in. R-7 extruded foam exterior insulation.
Vertical Glazing (Envelope)	U-factor 0.26 north and west windows, 28% window-wall ratio.
Window Design for Thermal Conditions	0.37 SHGC north and west windows, clerestory windows opened and closed by EMS as part of the natural ventilation cooling system.
Window Design for Daylight	Overhangs on east and south clerestories, south and north view glass for daylighting.
Lighting	
Daylighting	East and south clerestories, south and north windows, EMS stepped controlled area lights based on interior photocells, motion sensor-controlled lights in offices, hallways, break area, restrooms, and storeroom.
Electric Lighting Design	1.0 W/ft^2 in bookstore, T-8 with 88% uplight fixtures, 2-W LED exit signs, compact fluorescents, fluorescent and metal halide exterior lights controlled by EMS.
HVAC	
Equipment	All cooling provided with passive, direct evaporative cooling (cooltowers) and natural ventilation. All heating provided by electric radiant heating panels. No ductwork or mechanical room.
Ventilation	Natural ventilation through EMS-operated clerestory windows.
Controls	15 radiant zones separately controlled by EMS with wall-mounted temperature sensors.
Service Water Heating	
SWH	Three 1.6 kW storage electric hot water heaters with EMS bypass to limit use during peak demand events.
Additional Savings	
Plug Loads	EMS controls heating panels, and service hot water heaters limit peak demand charges.
Exterior Lighting	EMS controlled exterior lights for 2 to 4 hours after closing.
Other	7.2 kW of PV integrated with a UPS system.

CLIMATE ZONE 7—BIGHORN HOME IMPROVEMENT CENTER

SILVERTHORNE, COLORADO

The BigHorn Home Improvement Center in Silverthorne, Colorado, is located in climate zone 7. It consists of an 18,400 ft^2 retail hardware store and a 24,000 ft^2 building materials warehouse. The building was designed for its mountain climate, which is heating dominated with more than 10,000 (base 65°F) heating degree-days. Hourly computer energy simulations were used to design the energy features with a goal of minimizing energy costs. The building was monitored to ensure that design goals were met.

Energy-saving features include retail sales floor lighting designed to 1.1 W/ft^2, R-38 continuous roof insulation and R-19 wall insulation with R-12.5 or R-5 continuous exterior insulation, rigid c.i. beneath the entire floor to R-10 as well as the edges, glazing with a U-value of 0.24, motion sensors for restroom and office lighting, clerestory windows to provide daylighting, and daylighting controls to "dim" lights by turning off lamps (there are five steps of lighting levels). Plug loads, including vending machines, are only energized during business hours.

The building faces east, but the clerestory windows were designed to face north and south to best use the daylighting. The clerestories have overhangs to minimize the

Photograph courtesy of DOE/NREL

Figure 4-13. BigHorn Home Improvement Center entrance.

(a)

Photographs courtesy of DOE/NREL

(b)

Figure 4-14. (a) Daylighting controls dim lights when there is enough natural light. (b) Solar panels on the roof and clerestory windows.

summer solar gain. Even though the lights are installed to a density of 1.1 W/ft^2, the maximum recorded density has been 1.0 W/ft^2 due to daylighting controls. The daylighting, which provides 67% of the lighting in the retail area, reduces cooling loads to the point that conventional air conditioning is not needed, even though it is typically installed in retail buildings in this climate.

Additional Features

The store also has 9 kW of PVs integrated into the standing seam metal roof. Natural ventilation between the doors and the automatically controlled clerestory windows provides all the cooling loads for the space. Radiant floor heat is designed to provide extra heat to the occupied areas, including the cash registers. The designers decided on the hydronic radiant floor system for the retail/office area because of (1) the reduced noise from the elimination of the fans, (2) the improved comfort from the warm floor, and (3) the ability to have multiple zones within one large open space for better heating control. The energy management system controls the lights, natural ventilation, and heating system. The overall annual energy cost savings has been measured at 53% compared to a similar store built to meet the minimum requirements of ASHRAE Standard 90.1-2001.

The following how-to tips were implemented in this project: QA1, QA3, QA16, EN4, EN14, EN22, EN23, EN26, EN 27, DL1, DL2, DL3, DL7, DL9, EL13, EL14, EL15, EL16, EL19, PL1, PL2, and PL4.

BIGHORN HOME IMPROVEMENT CENTER	
Processes for Achieving Energy Savings	**Description of Project Elements**
Envelope	
Opaque Envelope Components	R-38 c.i. above deck insulation. R-19 insulated cavity between 24 in. o.c. steel studs, plus R-12.5 or R-5 c.i. for exterior insulation. R-10 c.i. under entire slab floor and vertically along footings.
Vertical Glazing (Envelope)	U-factor 0.24, 9.2% window-to-wall ratio.
Window Design for Thermal Conditions	0.44 SHGC, clerestory windows opened and closed by EMS as part of the natural ventilation system.
Window Design for Daylight	Overhangs on north and south clerestories.
Lighting	
Daylighting	North and south clerestories, high dormer windows, white walls, floor, and vaulted ceiling of the retail area, 5 levels stepped controls, EMS control based on interior photocells, occupancy sensors in restrooms and offices.
Electric Lighting Design	1.1 W/ft^2 pendant-type, fluorescent fixtures, 5 lighting levels to match daylighting availability.
HVAC	
Equipment	85% efficient gas boilers for radiant floor heating system, no air-conditioning system, all cooling provided with natural ventilation.
Ventilation	Natural ventilation through EMS-operated clerestory windows and open front door.
Controls	9 radiant zones separately controlled by EMS with wall and slab-mounted temperature sensors.
Service Water Heating	
SWH	Two 10 gal electric hot water heaters for the restrooms and service sinks.
Additional Savings	
Plug Loads	Vending machines disabled during off-hours.
Exterior Lighting	HID, EMS controlled by photocell for 2 hours after closing and 1 hour before opening.
Other	9 kW of building integrated PV, 400 W demonstration wind turbine attached to building.

How to Implement Recommendations

5

Recommendations are contained in the individual tables in Chapter 3, "Recommendations by Climate." The following how-to tips are intended to provide guidance on good practices for implementing the recommendations as well as cautions to avoid known problems in energy-efficient construction.

QUALITY ASSURANCE

Quality and performance are never an accident. They are always the result of high intention, sincere effort, intelligent direction, and skilled execution. A high-quality building that functions in accordance with its design intent, and thus meets the performance goals established for it, requires that quality assurance (QA) be an integral part of the design and construction process as well as the continued operation of the facility. This process is typically referred to as *commissioning*.

To reduce project risk, commissioning (Cx) requires a dedicated person (one with no other project responsibilities) who can execute a systematic process that verifies that the systems and assemblies perform as required. An independent party, whether it is a third-party Cx professional or a capable member of the organization of the installing contractor, architect, or engineer of record, is needed to ensure that the strategy sets and recommendations contained in this Guide meet the owner's stated requirements. This person is the commissioning authority, or CxA.

The Cx process defined by ASHRAE's Guideline 0 and Guideline 1 are applicable to all buildings. Owners, occupants, and the delivery team benefit equally from the QA process. Large and complex buildings require a correspondingly greater level of effort than that required for small, simpler buildings. Small retail buildings covered by this Guide have relatively simple systems and generally do not require the level of Cx effort required for more complex buildings. The following Cx practice recommendations meet this objective.

Activity	Complete
Owner selects CxA/QA provider and commitment to QA to designers and, through the contract documents, to contractors. The owner's responsibility includes directing the team to resolve issues identified through the QA process.	
CxA/QA provider reviews the Owner's Project Requirements and the designers' basis-of-design documentation for completeness and clarity and identifies areas requiring further clarification.	
CxA/QA provider conducts focused review of 100% construction documents that verifies the design meets the defined objectives and criteria established by Owner's Project Requirements and documents concerns to owner and designers.	
CxA/QA provider reviews comments from design review with designers and owner and adjudicates issues.	
CxA/QA provider develops Cx specifications that define team roles and responsibilities and pass/fail criteria for performance verification.	
CxA/QA provider assists design team by providing overview of process to prospective bidders and answers questions at pre-bid meeting.	
CxA/QA provider prepares construction checklists and Cx plan and conducts meeting with project team and establishes tentative schedule for Cx activities.	
CxA/QA provider reviews submittal information for systems being commissioned and develops functional test procedures contractors will use to demonstrate commissioned system performance.	
CxA/QA provider conducts two site visits during construction to verify that concerns identified during 100% construction document review were corrected and to identify issues that would affect performance.	
CxA/QA provider schedules testing through GC and directs, witnesses, and documents the functional testing that demonstrates performance.	
CxA/QA provider reviews O&M information and verifies that owner is trained in warranty and preventive maintenance requirements and has operational and maintenance information needed to meet the requirements.	

Note that the following how-to tips address the recommendations in Chapter 3, as they are generally applicable to many specific construction projects.

Good Design Practice

QA1   Select Team

Selection of the correct team members is critical to the success of a project. Owners who understand the connection between a building's performance and its impact on the environment, the psychological and physiological perceptions of occupants, and the total cost of ownership also understand the importance of team dynamics in selecting the team members responsible for delivering their project. Owners should evaluate qualifications of candidates, their past performance, the cost of their services, and their availability when making a selection. Once the team is selected, a pre-design meeting should be held to define team members' roles and responsibilities. This includes defining deliverables at each phase of the process and the Cx process.

QA2   Selection of Quality Assurance Provider

Quality assurance is a systematic process of verifying the Owner's Project Requirements, operational needs, and basis of design and ensuring that the building performs in accordance with these defined needs. The selection of a QA provider should include the same evaluation process the owner would use to select other team members. Qualifications in providing QA services, past performance of projects, cost of services, and availability of the candidate are some of the parameters an owner should investigate and consider when making a selection. Owners may select a member of the design or construction team as the QA provider. While there are exceptions, in general most designers are not comfortable operating and testing assemblies and equipment and most contractors do not have the technical background necessary to evaluate performance. Commissioning requires in-depth technical knowledge of the building envelope and the mechanical, electrical, and plumbing systems and operational and construction experience. This function is best performed by a third party responsible to the owner because political issues often inhibit a member of the design or construction organizations from fulfilling this responsibility.

QA3   Owner's Project Requirements

The Owner's Project Requirement (OPR) document details the functional requirements of a project and the expectations of how the facility will be used and operated. This includes strategies and recommendations selected from this Guide (see Table 2-5 and Chapters 3 and 4) that will be incorporated into the project, anticipated hours of operation provided by the owner, and basis-of-design assumptions. The OPR document forms the foundation of the team's tasks by defining project and design goals, measurable performance criteria, owner directives, budgets, schedules, and supporting information in a single, concise document. The QA process depends on a clear, concise, and comprehensive OPR document.

Development of the OPR document requires input from all key facility users and operators. The OPR document evolves through each project phase and contains documented decisions made during the design, construction, occupancy, and operation phases. This becomes the primary document for recording success and quality at all phases of the project delivery and throughout the life of the facility. Included in the OPR document are the designers' assumptions, which form the basis of design. The basis of design records the concepts, calculations, decisions, and product selections used to meet the OPRs and to satisfy applicable regulatory requirements, standards, and guidelines.

Note: The OPR document remains relatively fixed from its initial development until directed otherwise by the owner.

QA4   Budgets Contained in the OPR Document

The OPR document is used to define the team's scope in both broad and specific terms. It also defines the QA scope and budgets. The effort and cost associated with designing and constructing an energy-efficient building can be and often are lost because the performance of systems is not verified.

QA5   Design and Construction Schedule

The inclusion of QA activities in the construction schedule fulfills a critical part of delivering a successful project. Identify the activities and time required for design review and performance verification to minimize time and effort needed to accomplish activities and correct deficiencies.

QA6   Design Review

A second pair of eyes provided by the CxA/QA provider gives a fresh perspective that allows identification of issues and opportunities to improve the quality of the construction documents with verification that the OPRs are being met. Issues identified can be more easily corrected early in the project, providing potential savings in construction costs and reducing risk to the team. (See "Suggested Commissioning Scope" in Chapter 2 for more detail.)

QA7   *Defining Quality Assurance at Pre-Bid*

The building industry has traditionally delivered buildings without using a verification process. Changes in traditional design and construction procedures and practices require education of the construction team that explains how the QA process change will affect the various trades bidding the project. It is extremely important that the QA process be reviewed with the bidding contractors to facilitate understanding of and to help minimize fear associated with new practices. Teams who have participated in the Cx process typically appreciate the process because they are able to resolve problems while their manpower and materials are still on the project, significantly reducing delays, callbacks, and associated costs while enhancing their delivery capacity.

QA8   *Verifying Building Envelope Construction*

The building envelope is a key element of an energy-efficient design. Compromises in assembly performance are common and are caused by a variety of factors that can easily be avoided. Improper placement of insulation, improper sealing or lack of sealing around air barriers, incorrect or poorly performing glazing and fenestration systems, incorrect placement of shading devices, misplacement of daylighting shelves, and misinterpretation of assembly details can significantly compromise the energy performance of the building (see "Cautions" throughout this chapter). The perceived value of the Cx process is that it is an extension of the quality control processes of the designer and contractor as the team works together to produce quality energy-efficient projects.

QA8A   *Verifying Lighting Construction*

In small retail buildings, lighting plays a significant role in the energy consumption of the building; its impact becomes more pronounced in cooling-dominated climates. Lighting will often be designed after construction of the shell. If possible, the lighting loads should be specified before selection of the HVAC systems in order to select the size and system type for the most efficient and cost-effective approach.

QA9   *Verifying Electrical and HVAC Systems Construction*

Performance of electrical and HVAC systems are key elements of this Guide. How systems are installed affect how efficiently they can be serviced and how well they will perform. Observations during construction identify problems when they are easy to correct.

QA10   *Performance Testing*

Performance testing of systems is essential to ensuring that a project following this Guide will actually attain the energy savings that can be expected from the strategies and recommendations contained in this Guide (see "Suggested Commissioning Scope" in Chapter 2 for the CxA/QA provider responsibilities). If the contractors utilize the checklists as intended, functional testing of systems will occur quickly and minor but important issues will need to be resolved to ensure that the building will perform as intended. Owners with operational and maintenance personnel can use the functional testing process as a training tool to educate their staff on how the systems operate as well as for system orientation prior to training.

QA11   *Substantial Completion*

Substantial completion generally means the completion and acceptance of the life safety systems. Contractors, typically, do not test the systems' performance at substantial completion. While the systems may be operational, they probably are not yet operating as intended. Expected performance can only be accomplished when all systems operate interactively to provide the desired results. As contractors finish their work, they will identify and resolve many performance problems. The CxA/QA provider helps to resolve remaining issues.

QA12   Maintenance Manual Submitted and Accepted

The Cx/QA process includes communication of activities that the owner will be responsible for completing in order to maintain the manufacturers' warranties (see "Suggested Commissioning Scope" in Chapter 2 for QA provider responsibilities). A copy of the OPR document should be provided to the operation and maintenance (O&M) staff for an understanding of how the building is intended to operate.

QA13   Resolve Quality Control Issues Identified Throughout the Construction Phase

Issues identified during the construction process are documented in an "issues log" and presented to the team for collaborative resolution. Issues are tracked and reviewed at progress meetings until they are resolved. Typically the CxA develops and maintains the issues log. Completion and acceptance of the systems and assemblies by the owner will be contingent upon what issues are still outstanding at the end of the project. Minor issues may be tracked by the owner's O&M staff, while other issues will require resolution before acceptance of the work. The Cx/QA process finishes with verification that the issues identified have been resolved. The owner provides direction to the team to resolve issues identified.

QA14   Final Acceptance

Final acceptance generally occurs after the Cx/QA issues in the issues log have been resolved, except for minor issues the owner is comfortable with resolving during the warranty period.

QA15   Establish Building Maintenance Program

Continued performance and control of O&M costs require a maintenance program. The O&M manuals provide information that the O&M staff uses to develop this program. Detailed O&M system manual and training requirements are defined in the OPR document and executed by the project team to ensure O&M staff has the tools and skills necessary. The level of expertise typically associated with O&M staff for buildings covered by this Guide is generally much lower than that of a degreed or accredited engineer, and they typically need assistance with development of a preventive maintenance program. The CxA/QA provider can help bridge the knowledge gaps of the O&M staff and assist the owner with developing a program that would help ensure continued performance. The benefits associated with energy-efficient buildings are realized when systems perform as intended through proper design, construction, operation, and maintenance.

QA16   Monitor Post-Occupancy Performance

Establishing measurement and verification procedures with a performance baseline from actual building performance after it has been commissioned can identify when corrective action and/or repair is required to maintain energy performance. Utility consumption and factors affecting utility consumption should be monitored and recorded to establish building performance during the first year of operation.

Variations in utility usage can be justified based on changes in conditions typically affecting energy use, such as weather, occupancy, operational schedule, maintenance procedures, and equipment operations required by these conditions. While most buildings covered in this Guide will not use a formal measurement and verification process, tracking the specific parameters listed above does allow the owner to quickly review utility bills and changes in conditions. Poor performance is generally obvious to the reviewer when comparing the various parameters. CxA/QA providers can typically help owners understand when operational tolerances are exceeded and can provide assistance in defining what actions may be required to return the building to peak performance.

ENVELOPE

Opaque Envelope Components

Good Design Practice

EN1   *Cool Roofs* (Climate Zones: ❶ ❷ ❸)

Cool roofs are recommended for roofs with insulation entirely above deck and for metal building roofs. In order to be considered a cool roof for climate zones 1–3, a solar reflectance index (SRI) of 78 or higher is recommended, as determined by ASTM E 1980. The solar reflectance and thermal emmittance property values should represent long-term performance, such as three-year aged values to account for aging and soiling of roofs. Ratings should be determined by a laboratory accredited by the Cool Roof Rating Council.

EN2   ***Roofs, Insulation Entirely above Deck***

The insulation entirely above deck (see Figure 5-1) should be continuous insulation (c.i.) rigid boards because there are no framing members present that would introduce thermal bridges or short circuits to bypass the insulation.

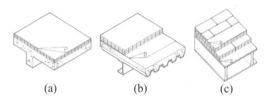

(a)   (b)   (c)

Figure 5-1. (EN2) Insulation entirely above deck—insulation is installed above (a) concrete, (b) metal, or (c) wood deck in a continuous manner.

When two layers of c.i. are used in this construction, the board edges should be staggered to reduce the potential for convection losses or thermal bridging. If an inverted or protected membrane roof system is used, at least one layer of insulation is placed above the membrane while a maximum of one layer is placed beneath the membrane.

EN3   ***Roofs, Metal Buildings***

In metal roof building construction, purlins are typically z-shaped cold-formed steel members, although steel bar joists are sometimes used for longer spans.

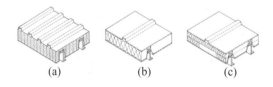

(a)   (b)   (c)

Figure 5-2. (EN3) Prefabricated metal roofs showing thermal blocking of purlins.

The thermal performance of metal building roofs with fiberglass blankets is improved by addressing the thermal bridging associated with compression at the purlins. The two types of metal building roofs are standing seam roofs and through-fastened roofs. Standing seam roofs have very few exposed fasteners and utilize a concealed clip for the structural attachment of the metal roof panel to the purlins. The larger gap between the purlin and the roof sheets, along with the thermal space block, provides a thermal break that results in improved performance compared to the standard through-fastened metal roofs. It is recommended that the thermal resistance between the purlin and the metal deck be at least R-8. One means to accomplish this is by using a 3/4 × 3 in. foam block (R-5) over 3/4 in. of compressed fiberglass blanket (R-3) (see Figure 5-2). Alternatively, a 2 in. space filled with compressed fiberglass insulation will provide roughly R-8.

Through-fastened metal roofs are screwed directly to the purlins and have fasteners that are exposed to the elements. The fasteners have integrated neoprene washers under the heads to provide a weathertight seal. Thermal spacer blocks are not used with through-fastened roofs because they may diminish the structural load-carrying capacity by "softening" the connection and restraint provided to the purlin by the metal roof panels. To meet the performance recommendations of this Guide, through-fastened roofs will generally require insulation over the purlins in the conventional manner, with a second layer of insulation added to the system. The second layer of insulation can be placed either parallel to the purlins (on top of the first layer) or suspended below the purlins.

In climate zones 1 and 2, the recommended construction is standing-seam roofs with R-19 insulation blankets draped over the purlins.

In climate zones 3 through 8, the recommended construction is standing-seam roofs with two layers of blanket insulation. The first layer is draped perpendicularly over the purlins with enough looseness to allow the second insulation layer to be laid above it, parallel to the purlins.

In any case, rigid c.i. or other high-performance insulation systems may be used to meet the U-factors listed in Appendix A.

EN4   *Roofs, Attics, and Other Roofs* (Climate Zones: all)

Attics and other roofs include roofs with insulation entirely below (inside of) the roof structure (i.e., attics and cathedral ceilings) and roofs with insulation both above and below the roof structure (see Figure 5-3). Ventilated attic spaces need to (a) have the insulation installed at the ceiling line. Unventilated attic spaces may have the insulation

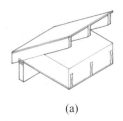 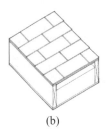

(a)   (b)

Figure 5-3. (EN4) Attics and other roofs.

installed at the roof line. When suspended ceilings with removable ceiling tiles are used, (b) the insulation needs to be installed at the roof line. For buildings with attic spaces, ventilation should be provided equal to 1 ft^2 of open area per 100 ft^2 of attic space. This will provide adequate ventilation as long as the openings are split between the bottom and top of the attic space.

EN5   *Roofs, Single Rafter* (Climate Zones: all)

Single rafter roofs have the roof above and ceiling below both attached to the same wood rafter, and the cavity insulation is located between the wood rafters (see Figure 5-4). Continuous insulation, when recommended, is installed to the bottom of the rafters and above the ceiling material. Single rafters can be constructed using solid wood framing members or truss type framing members. The cavity insulation should be installed

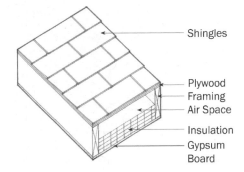

Shingles

Plywood
Framing
Air Space

Insulation
Gypsum
Board

Figure 5-4. (EN5) Wood joists, single rafter.

between the wood rafters and in intimate contact with the ceiling to avoid the potential thermal short-circuiting associated with open or exposed air spaces.

EN6   *Walls, Mass* (Climate Zones: all)

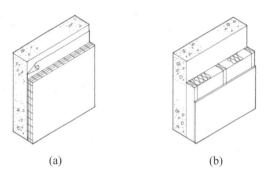

(a)   (b)

Figure 5-5. (EN6) Walls, mass—any concrete or masonry wall with a heat capacity exceeding 7 Btu/ft^2·°F.

Mass walls are defined as those with a heat capacity exceeding 7 Btu/ft^2·°F. Insulation may be placed on either the inside or the outside of the masonry wall. When insulation is placed (a) on the exterior of the wall, rigid c.i. is recommended (see Figure 5-5); when insulation is placed (b) on the interior of the wall, a furring or framing system may be used, provided the total wall assembly has a U-factor that is less than or equal to the appropriate climate zone construction listed in Appendix A.

The greatest advantages of mass can be obtained when insulation is placed on the exterior of the mass. In this case, the mass absorbs internal heat gains that are later released in the evenings when the buildings are not occupied.

EN7   *Walls, Metal Building* (Climate Zones: all)

In climate zones 1 and 2, a single layer of fiberglass batt insulation is recommended. The insulation is installed continuously perpendicular to the exterior of the girts and is compressed as the metal skin is attached to the girts (see Figure 5-6).

In climate zones 3 through 8, two layers of fiberglass batt insulation are recommended. The first layer is installed continuously perpendicular to the exterior of the girts and is compressed as the metal skin is attached to the girts. The second layer of insulation is installed parallel to the girts within the framing cavity.

In all climate zones, rigid c.i. is another option provided the total wall assembly has a U-factor that is less than or equal to the appropriate climate zone construction listed in Appendix A.

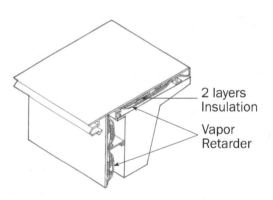

2 layers
Insulation

Vapor
Retarder

Figure 5-6. (EN7) Walls, metal building.

EN8 **Walls, Steel Framed (Climate Zones: all)**

Cold-formed steel framing members are thermal bridges to the cavity insulation (see Figure 5-7). Adding exterior foam sheathing as c.i. is the preferred method to upgrade the wall thermal performance because it will increase the overall wall thermal performance and tends to minimize the impact of the thermal bridging.

Alternative combinations of cavity insulation and sheathing in thicker steel-framed walls can be used provided that the proposed total wall assembly has a U-factor that is less than or equal to the

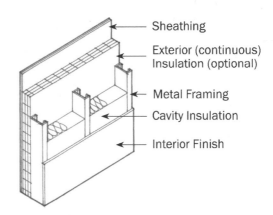

Sheathing

Exterior (continuous) Insulation (optional)

Metal Framing

Cavity Insulation

Interior Finish

Figure 5-7. (EN8) Walls, steel framed—a common construction type in nonresidential buildings.

U-factor for the appropriate climate zone construction listed in Appendix A. Batt insulation when installed in cold-formed steel-framed wall assemblies is to be ordered as "full width batts" and installation is normally by friction fit.

EN9 **Walls, Wood Frame and Other (Climate Zones: all)**

Cavity insulation is used within the wood-framed wall, while rigid c.i. is placed on the exterior side of the framing (see Figure 5-8). Care must be taken to have a vapor barrier on the warm side of the wall and to utilize a vapor-barrier-faced batt insulation product to avoid insulation sagging away from the vapor barrier.

Alternative combinations of cavity insulations and sheathings in thicker walls can be used provided the total wall assembly has a U-factor that is less than or equal to the appropriate climate zone construction listed in Appendix A.

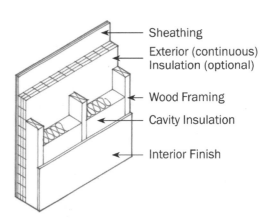

Sheathing

Exterior (continuous) Insulation (optional)

Wood Framing

Cavity Insulation

Interior Finish

Figure 5-8. (EN9) Walls, wood frame and other.

EN10   *Below-Grade Walls* (Climate Zones: all)

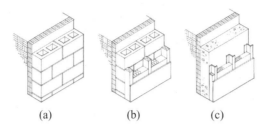

(a)   (b)   (c)

Figure 5-9. (EN10) Below-grade walls—the outer surface of the wall is in contact with the earth, and the inside surface is adjacent to conditioned or semi-heated space.

Insulation, when recommended, may be placed on either the inside or the outside of the below-grade wall (see Figure 5-9). If placed on the exterior of the wall, (a) rigid c.i. is recommended. If placed on the interior, a (b) furring or (c) framing system is recommended provided the total wall assembly has a C-factor that is less than or equal to the appropriate climate zone construction listed in Appendix A.

EN11   *Floors, Mass* (Climate Zones: all)

Insulation should be continuous and either integral to or above the slab (see Figure 5-10). It should be purchased by the conductive R-value. This can be achieved by (a) placing high-density extruded polystyrene as c.i. above the slab with either plywood or a thin layer of concrete on top. Placing insulation below the deck is not recommended due to losses through any concrete support columns or through the slab perimeter.

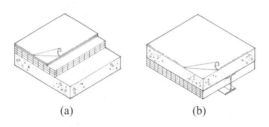

(a)   (b)

Figure 5-10. (EN11) Floors, mass—any floor with a heat capacity exceeding 7 Btu/ft^2·°F.

Exception: Buildings or zones within buildings that have durable floors for heavy machinery or equipment could have (b) insulation placed below the deck.

When heated slabs are placed below grade, below-grade walls should meet the insulation recommendations for perimeter insulation according to the heated slab-on-grade construction.

EN12   *Floors, Steel Joist or Wood Frame* (Climate Zones: all)

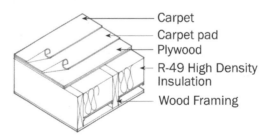

— Carpet
— Carpet pad
— Plywood
— R-49 High Density Insulation
— Wood Framing

Figure 5-11. (EN12) Floors, wood frame.

Insulation should be installed parallel to the framing members and in intimate contact with the flooring system supported by the framing member in order to avoid the potential thermal short-circuiting associated with open or exposed air spaces (see Figure 5-11).

Nonrigid insulation should be supported from below no less frequently than 24 in. on center.

EN13 *Slab-on-Grade Floors, Unheated* (Climate Zones: ➎ ➏ ➐ ➑)

As shown in Figure 5-12, (a) rigid c.i. should be used around the perimeter of the slab and should reach the depth listed in the recommendation or to the bottom of the footing, whichever is less; (b) additionally, in climate zones 5 through 8 and in cases where the frost line is deeper than the footing, c.i. should be placed beneath the slab as well.

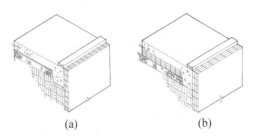

(a)   (b)

Figure 5-12. (EN13) Slab-on-grade floors, unheated—no heating elements either within or below the slab.

EN14 *Slab-on-Grade Floors, Heated* (Climate Zones: all)

When slabs are heated (see Figure 5-13), rigid c.i. should be used around the perimeter of the slab and should reach to the depth listed in the recommendation or to the bottom of the footing, whichever is less. Additionally, in climate zones 5 through 8, c.i. should be placed below the slab as well. Note that it is important to use the conductive R-value for the insulation as radiative heat transfer is small in this application.

Note: In areas where termites are a concern and rigid insulation is not recommended for use under the slab, a different heating system should be used.

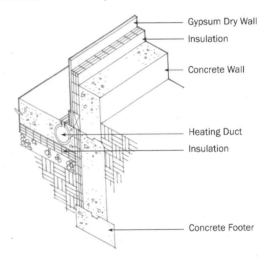

Gypsum Dry Wall

Insulation

Concrete Wall

Heating Duct
Insulation

Concrete Footer

Figure 5-13. (EN14) Slab-on-grade floors, heated—heating elements either within (as shown) or below the slab.

EN15 *Doors, Opaque and Swinging* (Climate Zones: all)

A U-factor of 0.37 corresponds to an insulated double-panel metal door. A U-factor of 0.61 corresponds to a double-panel metal door. If at all possible, single swinging doors should be used. Double swinging doors are difficult to seal at the center of the doors (see Figure 5-14) unless there is a center post. Double swinging doors without a center post should be minimized and limited to areas where width is important. Vestibules can be added to further improve the energy efficiency.

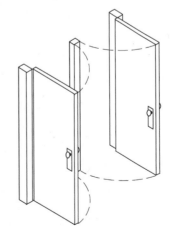

Figure 5-14. (EN15) Doors, swinging—opaque doors with hinges on one side.

EN16   ***Doors, Opaque and Roll-Up or Sliding*** **(Climate Zones: all)**

Roll-up or sliding doors are recommended to have R-4.75 rigid insulation or meet the recommended U-factor. When meeting the recommended U-factor, the thermal bridging at the door and section edges is to be included in the analysis. Roll-up doors that have solar exposure should be painted with a reflective paint (or high emmissivity) and/or should be shaded. Metal doors are a problem in that they typically have poor emmissivity and collect heat, which is transmitted through even the best insulated door, causing cooling loads and thermal comfort issues in the space.

If at all possible, use insulated panel doors over roll-up doors, as the insulation values can approach R-10 and provide a tighter seal to minimize infiltration.

Options

EN17   ***Alternative Constructions*** **(Climate Zones: all)**

The climate zone recommendations provide only one solution for upgrading the thermal performance of the envelope. Other constructions can be equally effective, but they are not provided in this Guide. Any alternative construction that is less than or equal to the U-factor, C-factor, or F-factor for the appropriate climate zone construction is equally acceptable. A table of U-factors, C-factors, and F-factors that corresponds to all of the recommendations is presented in Appendix A.

Procedures to calculate U-factors and C-factors are presented in *ASHRAE Handbook—Fundamentals*, and expanded U-factor, C-factor, and F-factor tables are presented in Standard 90.1, Appendix A.

Cautions

The design of building envelopes for durability, indoor environmental quality, and energy conservation should not create conditions of accelerated deterioration, reduced thermal performance, or problems associated with moisture and air infiltration. The following cautions should be incorporated into the design and construction of the building.

EN18   ***Heel Heights*** **(Climate Zones: all)**

When insulation levels are increased in attic spaces, the heel height should be raised to avoid or at least minimize the eave compression.

EN19   ***Slab Edge Insulation*** **(Climate Zones: all)**

Use of slab-edge insulation improves thermal performance, but problems can occur in regions of the country that have termites.

EN20   ***Moisture Control*** **(Climate Zones: all)**

Building envelope assemblies (see Figures 5-15a and 5-15b) should be designed to prevent wetting, high moisture content, liquid water intrusion, and condensation caused by diffusion of water vapor. See *2005 ASHRAE Handbook—Fundamentals*, Chapter 24.

EN21   ***Air Infiltration Control*** **(Climate Zones: all)**

The building envelope should be designed and constructed with a continuous air barrier system to control air leakage into or out of the conditioned space. An air barrier system should also be provided for interior separations between conditioned space and space designed to maintain temperature or humidity levels that differ from those in the conditioned space by more than 50% of the difference between the conditioned space and design ambient conditions. The air barrier system should have the following characteristics:

- It should be continuous, with all joints made airtight.
- Air barrier materials used in frame walls should have an air permeability not to exceed 0.004 cfm/ft^2 under a pressure differential of 0.3 in. water (1.57 lb/ft^2) when tested in accordance with ASTM E 2178.

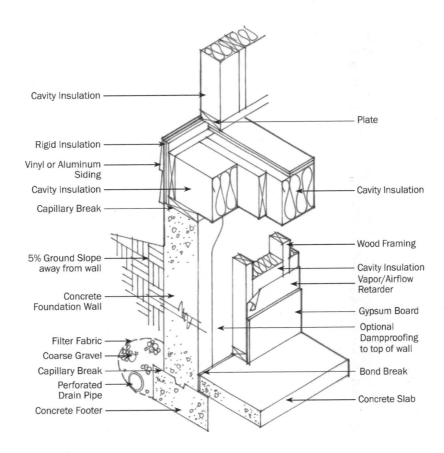

Figure 5-15a. (EN20) Moisture control for mixed climates.

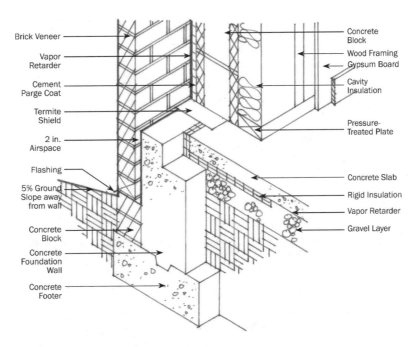

Figure 5-15b. (EN20) Moisture control for warm, humid climates.

- The system is capable of withstanding positive and negative combined design wind, fan, and stack pressures on the envelope without damage or displacement and should transfer the load to the structure. It should not displace adjacent materials under full load.
- It is durable or maintainable.
- The air barrier material of an envelope assembly should be joined in an airtight and flexible manner to the air barrier material of adjacent assemblies, allowing for the relative movement of these assemblies and components due to thermal and moisture variations, creep, and structural deflection.
- Connections should be made between:

     a. Foundation and walls
     b. Walls and windows or doors
     c. Different wall systems
     d. Wall and roof
     e. Wall and roof over unconditioned space
     f. Walls, floors, and roof across construction, control, and expansion joints
     g. Walls, floors, and roof to utility, pipe, and duct penetrations

- All penetrations of the air barrier system and paths of air infiltration/exfiltration should be made airtight.

Fenestration: Vertical Glazing and Skylights (Envelope)

Good Design Practice

EN22 **(Climate Zones: all)**

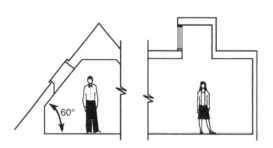

Figure 5-16. (EN22) Vertical fenestration—defined as slope greater than 60° from the horizontal.

The recommendations for fenestration are subdivided into those for vertical glazing (e.g., storefront windows, glazed doors, other windows) and those for skylights and are listed in Chapter 3 by climate zone. Vertical fenestration is defined as a slope greater than 60° from the horizontal (see Figure 5-16).

Table 5-1 lists the type of vertical glazing construction that generally corresponds to the U-factors and solar heat gain coefficient (SHGC) values in the Chapter 3 Recommendation Tables.

Table 5-1. Vertical Fenestration Descriptions

U-factor	SHGC	VLT	Class and Coating	Spacer	Frame
0.69	0.44	0.45	Double clear	Standard	Metal, vinyl, wood
0.49	0.40	0.45	Double tinted with low-e coating	Insulated	Metal with thermal break, vinyl, wood
0.41	0.41	0.60	Double clear with selective low-e coating	Standard	Metal with isolation bar, vinyl, wood
0.38	0.41	0.60	Double clear with selective low-e coating	Insulated	Metal with isolation bar, vinyl, wood

To be useful and consistent, the U-factors should be measured over the entire fenestration assembly, not just the center of glass. Look for a National Fenestration Rating Council (NFRC) label (or NFRC Label Certificate for site-built fenestration) that denotes the fenestration product is rated, certified, and labeled in accordance with NFRC procedures. Thermal performance of field-constructed fenestration systems should be verified using procedures described in AAMA 507-03 (revised April 2004). The selection of high-performance window products should be considered separately for each orientation of the building and for daylighting and viewing functions.

EN23 ***Vertical Glazing Area as a Percentage of Gross Wall Area***
(Climate Zones: all)

This is the percentage resulting from dividing the total fenestration, including glazed doors, by the gross exterior wall area. For any area less than 40%, the recommended values for U-factor and SHGC contribute to the 30% savings target of the entire building. A reduction in the overall fenestration area will also save energy, especially if glazing is significantly reduced on the east and west façades.

EN24 ***Skylight Area as a Percentage of Gross Roof Area*** **(Climate Zones: all)**

This is the percentage resulting from dividing the total skylight product area of the building by the gross roof area. Skylights provide increased daylight and a potential reduction in lighting energy consumption at the expense of increasing cooling loads in warmer climates and increasing heating loads in cooler climates. To achieve the lighting energy savings, the lighting in fixtures within ten feet of the skylight edge must have automatic controls that dim the lighting in response to available daylight. (See DL9 for guidance.)

EN25 ***Fenestration Design Guidelines for Thermal Conditions***

Uncontrolled solar heat gain is a major cause of energy consumption for cooling in warmer climates and thermal discomfort for occupants. Appropriate configuration of vertical glazing and skylights according to the orientation of the wall on which they are placed can significantly reduce these problems.

EN26 ***Solar Heat Gain Is Most Effectively Controlled on the Outside of the Building***
(Climate Zones: all)

Significantly greater energy savings are realized when sun penetration is blocked before entering the windows. Horizontal overhangs located at the top of the windows are most effective for south-facing façades, must continue beyond the width of the windows to adequately shade them, and need to be totally opaque. The vertical extension of the overhang depends on the height of the overhang from the bottom of the window sill (see Figure 5-17).

Note: Overhangs located directly above the window head need the least projection.

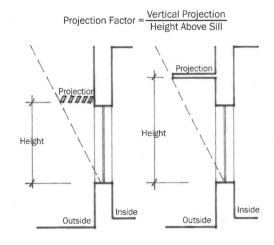

$$\text{Projection Factor} = \frac{\text{Vertical Projection}}{\text{Height Above Sill}}$$

Figure 5-17. (EN26) Windows with overhang.

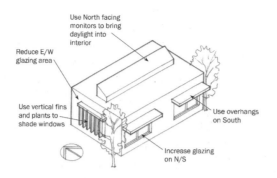

Figure 5-18. (EN26) Exterior sun control.

Vertical fins oriented slightly north are most effective for east- and west-facing façades (see Figure 5-18). Consider louvered or perforated sun control devices, especially in primarily overcast and colder climates, to prevent a totally dark appearance in those environments.

EN27 *Operable versus Fixed Windows* (Climate Zones: all)

Operable windows offer the advantage of personal comfort control and beneficial connections to the environment. However, individual operation of the windows not in coordination with the HVAC system settings and requirements can have impacts on the energy use of a building's system. Advanced energy buildings with operable windows should strive for a high level of integration between envelope and HVAC system design. First, the envelope should be designed to take advantage of natural ventilation with well-placed operable openings. Second, the mechanical system should employ interlocks on operable windows to ensure that the HVAC system responds by shutting down in the affected zone if the window is opened. It is important to design the window interlock zones to correspond as closely as possible to the HVAC zone affected by the open window.

In many retail spaces, large front doors can be opened and air relieved through dampers in the roof of the building. The relief dampers may be part of the HVAC system. In this case, natural ventilation can be achieved when exterior conditions warrant. The HVAC system should be turned off during such times. Many retailers like to open the doors and extend the retail environment to the sidewalk. This is an ideal time to deploy natural ventilation strategies.

Many smaller retail operations can benefit from operable windows to provide comfort control and natural ventilation. Like the previous example, the HVAC system should be turned off when natural ventilation is used.

EN28 *Building Form and Window Orientation* (Climate Zones: all)

In all climates, north- and south-facing glass can be more easily shielded and can result in less solar heat gain and less glare than can east- and west-facing glass. During site selection, preference should be given to sites that permit elongating the building in the east-west direction and that permit orienting more windows to the north and south (see Figure 5-19).

A good design strategy incorporates glass that contributes to the daylighting of the space. If possible, configure the building to maximize north-facing walls and glass by

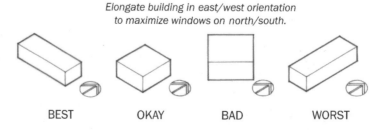

Figure 5-19. (EN28) Building and window orientation.

elongating the floor plan. Since sun control devices are less effective on the east and west façades, the solar penetration through the east- and west-facing glazing can cause a problem with glare and is usually shaded in the retail environment. This can be done by reducing the area of glazing, reducing the SHGC, or preferably both. Thus, the area of glazing on the east and west façades times their respective SHGCs should be less than the area of glazing on the north and south façades times their respective SHGCs. If each façade has a different area or SHGC, the formula becomes: ((W window area × W SHGC) + (E window area × E SHGC)) < ((N window area × N SHGC) + (S window area × S SHGC)). For buildings where a predominantly east-west exposure is unavoidable, or if the application of this equation would result in SHGCs of less than 0.25, then more aggressive energy conservation measures may be required in other building components to achieve an overall 30% energy savings.

Warm Climates

EN29 *Glazing* (Climate Zones: ❶ ❷ ❸ ○)

For north- and south-facing windows, select windows with a low SHGC and an appropriate visible light transmission. Certain window coatings, called *selective low-e*, transmit the visible portions of the solar spectrum selectively, rejecting the nonvisible infrared sections. These glass and coating selections provide superior view and daylighting while minimizing solar heat gain. Window manufacturers market special "solar low-e" windows for warm climates. For buildings in warm climates that do not utilize daylight-responsive lighting controls, glazing should be selected with a SHGC of no more than 0.44. All values are for the entire fenestration assembly, in compliance with NFRC procedures, and are not simply center-of-glass values. For warm climates, a low SHGC is much more important for low building energy consumption than the window assembly U-factor. Windows with low SHGC values will tend to have a low center-of-glass U-factor, however, because they are designed to reduce the conduction of the solar heat gain absorbed on the outer light of glass through to the inside of the window.

EN30 *Glazing* (Climate Zones: ❺ ❻ ❼ ❽)

For more northerly locations, only the south-facing glass receives much sunlight during the cold winter months. If possible, maximize south-facing windows by elongating the floor plan in the east-west direction and relocate windows to the south face. Be careful to install blinds or other sun-control devices for south-facing glass to allow for passive effects when desired but prevent unwanted glare and solar overheating. Glass facing east and west should be significantly limited. Areas of glazing facing north should be cautiously sized for daylighting and view. During site selection, preference should be given to sites that permit elongating the building in the east-west direction and that permit orienting more windows to the south. See also DL5.

Although higher SHGCs are allowed in colder climate zones, continuous horizontal overhangs are still useful for blocking summer sun. Window manufacturers market low-e glazing with higher SHGCs for cold climates.

EN31 *Obstructions and Planting* (Climate Zones: all)

Adjacent taller buildings and trees, shrubs, or other plantings are effective for shading glass on south, east, and west facades. For south-facing windows, remember that the sun is higher in the sky during the summer, so shading plants should be located high above the windows to effectively shade the glass. The glazing of fully shaded windows can be selected with higher SHGC ratings without increasing energy consumption. The solar reflections from adjacent buildings with reflective surfaces (metal, windows, or especially reflective curtain walls) should be considered in the design. Such reflections may modify shading strategies, especially on the north façade.

EN32   *Passive Solar* (Climate Zones: all)

Passive solar energy-saving strategies should be limited to non-sales and non-office spaces, such as lobbies and circulation areas, unless these strategies are designed so that workers and customers do not directly view interior sun patches or see them reflected on merchandise or work surfaces. Consider heat-absorbing blinds in cold climates or reflective blinds in warm climates. In spaces where glare is not an issue, the usefulness of the solar heat gain collected by windows can be increased by using massive thermally conductive floor surfaces, such as tile or concrete, in locations where the transmitted sunlight will fall. These floor surfaces absorb the transmitted solar heat gain and release it slowly over time, to provide a more gradual heating of the structure. Consider low-e glazing with exterior overhangs.

References

AAMA. 2003. *AAMA 507, Standard Practice for Determining the Thermal Performance Characteristics of Fenestration Systems Installed in Commercial Buildings*. Schaumburg, IL: American Architectural Manufacturers Association.

ASHRAE. 2005. *ASHRAE Handbook—Fundamentals*. Chapter 24, Thermal and moisture control in insulated assemblies—Applications. Atlanta: American Society of Heating, Refrigerating and Air-Conditioning Engineers, Inc.

ASTM. 2001. *ASTM E 1980, Standard Practice for Calculating Solar Reflectance Index of Horizontal and Low-Sloped Opaque Surfaces*. West Conshokocken, PA: American Society for Testing and Materials.

ASTM. 2003. *ASTM E 2178, Standard Test Method for Air Permeance of Building Materials*. West Conshokocken, PA: American Society for Testing and Materials.

IESNA. 1997. *EPRI Daylight Design: Smart and Simple*. New York: Illuminating Engineering Society of North America.

LBNL. 1997. Tips for daylighting with windows. *Windows & Daylighting*. Berkeley, CA: Lawrence Berkeley National Laboratories. http://windows.lbl.gov/daylighting/designguide/designguide.html.

Evans, Benjamin. 1997. *Daylighting Design, Time Saver Standards for Architectural Design Data*. New York: McGraw-Hill.

LIGHTING

Daylighting

Good Design Practice

DL1   *Savings and Occupant Acceptance* (Climate Zones: all)

Daylight in buildings can save energy if the electric lighting is switched or dimmed in response to changes in daylight levels in the store. Automatic lighting controls increase the probability that daylighting will save energy. It is also important that heat gain and loss through glazing be controlled. In addition, glare and contrast must be controlled so occupants are comfortable and will not override electric lighting controls. See additional comments related to skylight design and placement (EN23 and EN24).

DL2   *Surface Reflectance* (Climate Zones: all)

The use of light-colored materials and matte finishes in all daylighted spaces increases efficiency through interreflections and greatly increases visual comfort. See EL5.

DL3 *Control of Direct Sun Penetration* **(Climate Zones: all)**

Daylighting utilizes light from the sky, not the direct sun. Patches of direct sunlight in the sales area will create unacceptable brightness and excessive contrast between light and dark areas.

- Use exterior and interior sun control devices. Exterior sun control and overhangs help reduce both direct sun penetration and heat gain from vertical glazing surfaces.
- Use continuous exterior overhangs and interior horizontal blinds or shades on south-facing glazing.
- Use interior vertical slat blinds or shades on east- and west-facing glazing and as required for northeast or northwest façades.
- An exterior overhang needs to be deep enough to shield windows above the light shelf (if used) from direct sun. The light shelf, or the overhang if the light shelf is not used, should also be deep enough to shield windows below the shelf from direct sun.
- For "toplighting," use north-facing clerestories to avoid direct sun.
- For skylights, use light-reflecting baffles and/or diffusing glazing to control direct sun. Note that diffusing skylights can cause glare when the sun hits them.

DL4 *Skylight Thermal Transmittance* **(Climate Zones: all)**

Hot Climates

- Use north-facing clerestories for skylighting whenever possible in hot climates to eliminate excessive solar heat gain and glare. Typically, north-facing clerestories have one-sixth the heat gain of skylights.
- Reduce thermal gain during the cooling season by using skylights with a low over-all thermal transmittance (U-factor). Insulate the skylight curb above the roofline with rigid c.i.
- Shade skylights with exterior/interior sun control such as screens, baffles, or fins. See DL3.
- Use smaller aperture skylights in a grid pattern to gain maximum usable daylight with the least thermal heat transfer.

Moderate and Cooler Climates

- Use either north- or south-facing clerestories for skylighting but not east- or west-facing ones. East-west glazing adds excessive summer heat gain and makes it difficult to control direct solar gain. Clerestories with operable glazing may also help provide natural ventilation in temperate seasons when air conditioning is not in use. Typically, north-facing clerestories have one-sixth the heat gain of skylights.
- Reduce summer heat gain as well as winter heat loss by using skylights with a low overall thermal transmittance. Use a skylight frame that has a thermal break to prevent excessive heat loss/gain and winter moisture condensation on the frame. Insulate the skylight curb above the roofline with rigid c.i.
- Shade south-facing clerestories and skylights with exterior/interior sun control such as screens, baffles, or fins. See DL3.
- Use skylights with smaller apertures in a grid pattern to gain maximum usable daylight with the least thermal heat transfer. Do not exceed maximum prescribed glazing area.
- Splay skylight opening at 45° to maximize daylight distribution and minimize glare. See Figure 5-20.

DL5   Interactions (Climate Zones: all)

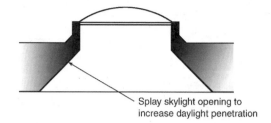

Thermal gains and losses associated with skylights should be balanced with daylight-related savings achieved by reducing electric lighting consumption.

Splay skylight opening at 45° to maximize daylight distribution and minimize glare. See Figure 5-20.

Figure 5-20. (DL4 and DL5) Splayed skylight opening.

DL6   Expanded Recommendations for Electric Lighting Controls in Daylight Zone (Climate Zones: all)

The *daylight zone* is the area of the skylight plus the floor-to-ceiling dimension in all directions from the edge of the skylight. The daylight zone at the perimeter is 15 ft deep and 2 feet wider than the window (see Figure 5-21).

Dimming controls. In merchandise sales areas, continuously dim rather than switch electric lights in response to daylight to minimize customer/employee distraction. Specify dimming ballasts that dim down to at least 20% of full output. Automatic multilevel daylight switching may be used in non-sales environments such as hallways, storage areas, restrooms, lounges, etc. To maintain an adaptation zone (high light level during the daylight hours), dimming of the luminaires adjacent to the entry is not recommended. Control luminaires in groups around skylights, and if using a lighting system that provides an indirect component, do not dim below 20% to maintain a brightness balance between skylights and surrounding ceiling. If daylight zones overlap, a single control zone may be used. The daylighting control system and/or photosensor should include a five-minute time delay or other means to avoid cycling caused by rapidly changing sky conditions and a one-minute fade rate to change the light levels by dimming.

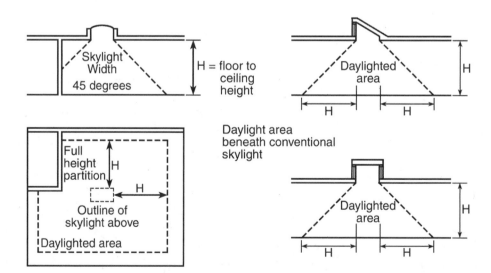

Figure 5-21. (DL6) Daylight zone.

DL7 ***Photosensor Placement* (Climate Zones: all)**

Photosensors used should be specified for the appropriate illuminance range (indoor) and must achieve a slow, smooth, linear dimming response from the dimming ballasts.

A *closed loop* system is one in which the interior photocell responds to the combination of daylight and electric light in the primary sales area. The best location for the photocell is above an unobstructed location such as a circulation path. If using a lighting system that provides an indirect component, mount the photosensor at the same height as the luminaire or in a location that is not impacted by the uplight from the luminaire.

An *open loop* system is one in which the photocell responds only to daylight levels but is still calibrated to the desired light level received on the merchandise. The best location for the photo sensor is inside the skylight well.

DL8 ***Calibration and Commissioning* (Climate Zones: all)**

Even a few days of occupancy with poorly calibrated controls can lead to permanent overriding of the system and loss of all savings. All lighting controls must be calibrated and commissioned after the finishes are completed and the merchandise is in place. Most photosensors require daytime and nighttime calibration sessions. The photosensor manufacturer and the QA provider should be involved in the calibration. Document the calibration and Cx settings and calendar intervals for future recalibration.

DL9 ***Daylight Levels* (Climate Zones: all)**

Occupants expect higher combined light levels in daylighted spaces. Consequently, it is more acceptable to occupants when the electric lights are calibrated to dim when the combined daylight and electric light on the merchandise exceeds 1.20 times the designed light level; i.e., if the ambient electric light level is designed for 50 maintained footcandles, the electric lights should begin to dim when the combined level is 60 footcandles ($50 \times 1.20 = 60$).

DL10 ***Interactions* (Climate Zones: all)**

Energy savings due to reduced electrical consumption should be weighed against any potential loss caused by increased cooling or heating loads.

References

IESNA. 1997. *EPRI Daylighting Design: Smart and Simple*. New York: Illuminating Engineering Society of North America.

IESNA. 1996. *EPRI Lighting Controls—Patterns for Design*. New York: Illuminating Engineering Society of North America.

NBI. 2003. *Advanced Lighting Guidelines*. White Salmon, WA: New Buildings Institute. www.newbuildings.org/lighting.htm.

USGBC. 2005. LEED NC Indoor Environment Quality Credit 6.1 "Controllability of Systems: Lighting." Washington, DC: U.S. Green Building Council.

Electric Lighting Design

Interior Lighting

Good Design Practice

Goals for Merchandise Lighting. The primary lighting goals common to all types of merchandising spaces are attracting and guiding customers, facilitating merchandise evaluation to initiate purchases, and enabling completion of the sale. Generally, merchandising lighting systems consist of three basic components:

Ambient lighting—The ambient lighting system should provide a general, uniform illuminance over the entire merchandising area or in contained areas of merchandising. This illuminance level may range from 15 to 75 footcandles depending on the store type and merchandising strategies. Designing to lower ambient footcandle levels can permit more effective accent lighting strategies.

Perimeter lighting—Perimeter lighting is important in providing impressions of a pleasant space, will help define a sense of space, and can assist in drawing customer attention to specific merchandising areas.

Accent lighting—Accent lighting provides areas of more intense illuminance, creating visual contrast and interest. Accent lighting can be used to create focus, define customer movement, and allow for more intense scrutiny of merchandise color, texture, and detail.

Other lighting features to consider that play important roles in merchandise recognition and the perceived lighted environment include decorative lighting (pendants, sconces, table lamps, etc.), internal casework lighting, and furniture/wall integrated lighting.

EL1 *Lighting Walls/Perimeter Lighting* (Climate Zones: all)

Better eye adaptation, luminous comfort, and impressions of pleasantness and space can be achieved when light is distributed to the walls. To light walls, use wall wash luminaires or locate ambient lighting fixtures closer to walls. There will be occurrences where it will be desirable to use accent lighting for certain perimeter wall features. When placing fixtures, always consider the final location of the merchandise to be illuminated, not the wall surface behind the merchandise (See Figure 5-22).

EL2 *Additional Interior Lighting/Accent Lighting* (Climate Zones: all)

The following additional lighting power densities (LPDs), from the Recommendation Tables in Chapter 3, are available for adjustable lighting equipment that is specifically designed and directed to highlight merchandise (accent lighting) above and beyond the base 1.3 W/ft^2 allowance. See EL12 for switching recommendations.

0.4 W/ft^2 (spaces not listed below)
0.6 W/ft^2 (sporting goods, small electronics)
0.9 W/ft^2 (furniture, clothing, cosmetics, and artwork)
1.5 W/ft^2 (jewelry, crystal, china)

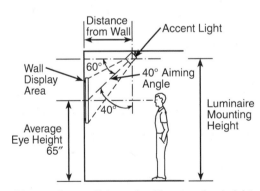

Distance from wall determined by mounting height, eye height, and 40° aiming angle.

Figure 5-22. (EL1 and EL2) Accent lighting aimed at 40°.

The use of accent lighting to highlight ALL merchandise does not create the proper contrast ratios. Use accent lighting to highlight key merchandise locations or vignettes to "feature display" light levels (three to ten times the general merchandise lighting level in the area of the display).

Along perimeters, accent lighting should be aimed up at 40° from vertical (straight down) to reduce reflected glare off specular surfaces. The aimed accent light should not exceed 45° from vertical, and attention should be given to both the direct and reflected characteristics of the light distribution. In areas open to customer view from multiple vantage points, the accent lighting should be aimed up at no more than 30° from vertical to reduce the possibility of direct glare visible to the customer (see Figure 5-22).

EL3 **Decorative Lighting (Climate Zones: all)**

Decorative lighting (wall sconces and pendant fixtures) can add visual interest and focus to the space, especially at the sales transaction area. These fixtures are included in the tabulation of the base LPD, and consideration should be given to energy-efficient solutions, including compact fluorescent, ceramic metal halide (CMH), and light-emitting diode (LED) lighting.

EL4 **Casework Lighting (Climate Zones: all)**

Casework lighting is not included in the tabulation of the LPD as long as it is integrated into the casework and is installed by the casework manufacturer. Lighting for casework must remain sensitive to the energy goals of the space. Strong consideration should be given to energy-efficient or low-energy solutions, including linear fluorescent, fiber optic, and linear LED sources.

EL5 **Reflectance (Climate Zones: all)**

Higher surface reflectance on ceilings and walls may increase store visibility through the front windows and will increase lighting levels and energy performance within the space. However, higher reflectance may not conform to a retailer's specific image. Energy savings outlined in this Guide are based on reflectance of 80-50-20 (ceiling-wall-floor). If the reflectance is lower, then additional attention to the ambient lighting energy requirements may be necessary. Avoid direct lighting of specular surfaces (mirrors, glass, polished metals, or polished stone) in customer areas, if possible; otherwise, consider carefully the reflected light component and its effect on the customer.

An 80%+ ceiling and 70%+ wall reflectance is preferred in daylight zones (see DL2). Reflectance values are available from paint and fabric manufacturers. Reflectance should be verified by the QA provider.

EL6 **Color Rendering Index (Climate Zones: all)**

The color rendering index (CRI) is a scale measurement identifying a lamp's ability, generally, to adequately reveal color characteristics. The scale maximizes at 100, with 100 indicating the best color-rendering capability. It is recommended that lamps specified for the ambient and accent lighting of retail merchandise have a CRI of 80 or greater to allow the consumer to effectively examine the color component of a product.

EL7 **Linear Fluorescent Lamps and Ballasts (Climate Zones: all)**

To achieve the LPD recommendations in Chapter 3, high-performance T-8 lamps and high-performance electronic ballasts are used for general lighting. The use of standard T-8 and energy-saving T-8 lamps may also be considered but may result in lower ambient light levels or an increased number of fixtures or lamps to achieve recommended light levels. Standard T-8 and energy-saving T-8 lamps also are available with lower CRI values than recommended in EL6, which may compromise the lighting solution.

High-performance T-8 lamps are defined, for the purpose of this Guide, as having a lamp efficacy of 90+ nominal lumens per watt, based on mean lumens divided by the cataloged lamp input watts. Mean lumens are published in the lamp catalogs as the degraded lumen output occurring at 40% of the lamp's rated life. High-performance T-8s also are defined as having a CRI of 81 or higher and 94% lumen maintenance. The higher performance is achieved either by increasing the output (3100 lumens) while keeping the same 32 W input as standard T-8s or by reducing the wattage while keeping the light output similar to standard T-8s (e.g., 2750 lumens for 28 W or 2850 lumens for 30 W).

High-performance electronic ballasts are defined, for the purpose of this Guide, as two-lamp ballasts using 55 W or less with a ballast factor (BF) of 0.87 or greater. One-lamp, three-lamp and four-lamp ballasts may be used but should have the same or better efficiency as the two-lamp ballast. Dimming ballasts do not need to meet this requirement. The higher output 3100 lumen lamps are visibly brighter than standard T-8s. Using ballasts with a BF of 0.77 may provide more comfortable lamp brightness, in direct luminaires where the lamp is visible, without sacrificing efficiency.

- *Program start* ballasts are recommended on frequently switched lamps (switched on and off more than five times a day) because they greatly extend lamp life over frequently switched instant start ballasts.
- *Instant start* T-8 ballasts typically provide greater energy savings and are the least costly option; also, the parallel operation allows one lamp to operate even if the other burns out. However, instant start ballasts may reduce lamp life, especially when controlled by occupancy sensors or daylight switching systems.
- T-5 ballasts should always be program start.

EL8   *Fluorescent T-5 Sources* (Climate Zones: all)

To achieve the LPD recommendations in Chapter 3, T-5HO and T-5 lamps may be part of the solution. They have initial lumens per watt that compare favorably to the high-performance T-8s. In addition to energy, T-5s use fewer natural resources (glass, metal, phosphors) than comparable lumen output T-8 systems. However, when evaluating the lamp and ballast as a system (at the mean lumens of the lamps), high-performance T-8 systems perform better than T-5HO systems. In addition, T-5s have higher surface brightness and should not be used in open-bottom fixtures. It may be possible to achieve the base LPD while maintaining the desired light levels using T-5 fixtures as the primary light source if careful selection of the fixture reduces direct glare from the lamp.

EL9   *Compact Fluorescent* (Climate Zones: all)

To achieve the LPD recommendations in Chapter 3, compact fluorescent lamps may be used for general ambient lighting and wall-washing. *Compact fluorescent lamps* are defined, for the purpose of this Guide, as having a lamp efficacy of 55+ nominal lumens per watt, based on mean lumens divided by the cataloged lamp input watts, and a CRI of 82 or greater. Use electronic ballasts on all compact fluorescent lighting.

Compact fluorescent lighting may be used for general store lighting but is less efficient than linear fluorescent and is much more expensive to dim. Compact fluorescent lighting should not be used for accent lighting.

EL10   *Ceramic Metal Halide* (Climate Zones: all)

To achieve the LPD recommendations in Chapter 3, ceramic metal halide (CMH) lamps may be used for general ambient, accent lighting, and wall-washing. *CMH lamps* are defined, for the purpose of this Guide, as having a lamp efficacy of 50+ nominal mean lumens per watt and a CRI of 81 or greater. Use electronic ballasts on all CMH lighting.

EL11   *Halogen IR* (Climate Zones: all)

To achieve the LPD recommendations in Chapter 3, halogen IR lamps may be used for accent lighting. *Halogen IR lamps* are defined, for the purpose of this Guide, as having a lamp efficacy of 20+ lumens per watt and using thin films to help redirect thermal energy through the filament, increasing light output.

CMH lamps can be up to 250% more efficient and have three to five times greater lamp life than halogen sources and are recommended over halogen IR lamps. However, there will likely be instances where the use of CMH lamps will not be practical as a result of cost considerations or because the size or ambient condition of the space will not support the intensity of CMH lamps. In these scenarios it may be necessary to consider the use of halogen lamps in either low voltage or line voltage forms. If halogen IR lamps are used, then it is recommended that the lamp envelopes be limited to MR16 bulb sizes. In all cases, the lamp wattage would be comparable to the lower wattage

CMH lamps; wattage is limited by the lamp size. PAR30IR and PAR38IR lamps could be used but are not recommended because of the breadth of lamp wattages and types available with which a fixture could be re-lamped at a future time. Standard incandescent and halogen sources are not recommended.

EL12   *Light-Emitting Diodes (Climate Zones: all)*

To achieve the LPD recommendations in Chapter 3, light-emitting diodes (LEDs) may be used for accent lighting. LEDs are solid-state semiconductor devices that can produce a wide range of saturated colored light and can be manipulated with color mixing or phosphors to produce white light.

White light LED sources should be carefully evaluated for use in the lighting of retail merchandise, as color rendering and color temperature capabilities can vary widely by manufacturer. Products are now available that allow the adjustment of the intensity and the white color temperature, providing dynamic flexibility to the retail environment. They are not particularly efficient at this time, with efficacy similar to that of halogen lamps, although the technology is rapidly improving.

LEDs are not well suited for the ambient lighting requirements of most spaces, but they can be used effectively for casework and display integrated lighting. Effective accent and wall-washing strategies can be achieved using LEDs, but intensity, color, and efficacy must be reviewed thoroughly. Single-color LEDs are well suited to create interesting visual effects, as they produce color more efficiently than filtering other white light sources with relatively low energy consumption.

Careful consideration should also be given to maintenance issues. LED lamp life can offer advantages over other sources but does vary by color. In many cases, the LED cannot be replaced as a single lamp, like an incandescent source, for example; often the entire control board will have to be removed and replaced.

EL13   *Occupancy Sensors* (Climate Zones: all)

Use occupancy sensors in all non-sales areas. The greatest energy savings are achieved with manual on/automatic off occupancy sensors if daylight is present. This avoids unnecessary operation when electric lights are not needed and greatly reduces the frequency of switching. In non-daylighted areas, ceiling-mounted occupancy sensors are preferred. In every application it should not be possible for the occupant to override the automatic OFF setting, even if set for manual ON. Unless otherwise recommended, factory-set occupancy sensors should be set for medium to high sensitivity and a 15-minute time delay (the optimum time to achieve energy savings without excessive loss of lamp life). Review manufacturer's data for proper placement and coverage.

The two primary types of occupancy sensors are *infrared* and *ultrasonic*. Infrared sensors can only see in a line-of-sight and should not be used in rooms where the user cannot see the sensor (e.g., storage areas with multiple aisles, restrooms with stalls). Ultrasonic sensors can be disrupted by high airflow and should not be used near air duct outlets.

EL14   *Lighting Circuits and Automatic Controls* (Climate Zones: all)

Put all general, all accent, and all display case lighting on separate circuits and switches (use multiple circuits and switches as required). Use automatic time scheduling (time switches) to turn on accent and display case lighting no more than 20 minutes prior to normal scheduled hours and to turn off accent and display case lighting no more than 20 minutes after normal scheduled hours.

It is also recommended, if track lighting is to be used, that the track be included with a current-limiting device. If installed, the limiting device can limit future additions of track heads and limit the amount of "assumed" power required for the LPD calculation.

EL15   *Electric Lighting Controls in Daylight Zones* **(Climate Zones: all)**

Factory setting of calibrations should be specified when feasible to avoid field labor. Lighting calibration and Cx should be performed after furniture installation but prior to occupancy to ensure user acceptance. Refer to DL1 through DL 10.

EL16   *Exit Signs* **(Climate Zones: all)**

Use LED exit signs or other sources that consume no more than 5 W per face. The selected exit sign and source should provide the proper luminance to meet all building and fire code requirements.

EL17   *Light Fixture Distribution* **(Climate Zones: all)**

Extensive use of totally indirect luminaires or recessed direct-indirect (coffer-type) fixtures may not achieve desired light levels while meeting the LPD goal from Chapter 3. Such fixtures can create inherent brightness/contrast problems and are not recommended.

EL18   *Overhead Glare Control* **(Climate Zones: all)**

Specify luminaires properly shielded for customer comfort. Avoid T-5 lamps in open-bottomed fixtures. Avoid highly specular louvers, cones, or reflectors visible to occupants from any angle. Use efficient fixtures and proper distribution. Use more fixtures of lower wattage rather than the reverse.

Sample Design Layouts for Retail Buildings

The 1.3 W/ft^2 recommendation for lighting power (shown in each Recommendation Table in Chapter 3) represents an average LPD for the entire building. Individual spaces may have higher power densities if they are offset by lower power densities in other areas. The example design described below is one way (but not the only way) that this watts-per-square-foot limit can be met.

EL19   *General Lighting in Merchandise Sales Areas* **(Climate Zones: all)**

The general lighting (fixture type G1 in Figure 5-23) at 1.5 W/ft^2 provides the base level of lighting for the merchandise. The spill light from the merchandise will provide adequate lighting for the circulation paths. Also included in the base allowances is decorative/focus lighting at the sales transaction counter—this may be provided by pendants over (fixture type P1 in Figure 5-23) and wall lighting behind (fixture type A2 in Figure 5-23) the counter.

General lighting can be provided by a number of different types fixtures.

- Direct fixtures, open, lensed or parabolic, designed as shown in Figure 5-23, will provide the highest footcandles for the space and will provide some positive shadowing on the product. However, if the displays are expected to change sometime after the installation, the fixtures may be misaligned to the displays. Consider an alternative layout with the fixtures running perpendicular to the displays.
- Indirect fixtures, pendant and recessed indirect, designed similarly to those shown in Figure 5-23, will not be display dependant but may look better running perpendicular to the display aisles. Indirect fixtures tend to provide a flat lighting effect and can draw the customers' attention away from the display.

EL20   *Accent Lighting in Merchandise Sales Areas* **(Climate Zones: all)**

Follow recommendation EL2 for accent lighting LPD above the base power allowance. Use accent lighting (fixture type A1 in Figure 5-23) to highlight key merchandise locations or vignettes to "feature display" light levels (three to ten times the general merchandise lighting level in the area of the display). The use of accent lighting to highlight ALL merchandise does not create the proper contrast ratios.

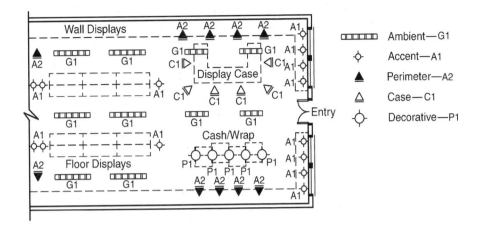

Figure 5-23. (EL19 through EL22) Layout for lighting in merchandise sales area.

Highlight window displays (fixture type A1 in Figure 5-23) to ten times the general merchandise lighting level to attract customers and to compete against potential high daylight levels. Window display lighting should be switched down to three times the general merchandise lighting level at night to help with eye adaptation when entering and exiting the store.

EL21 *Perimeter Lighting in Merchandise Sales Areas* (Climate Zones: all)

Follow recommendation in EL2 for perimeter lighting LPD above the base power allowance. Use accent or wall-washers (fixture type A2 in Figure 5-23) to highlight key wall locations to general merchandise lighting levels. It is especially important to highlight the back wall to draw customers' attention all the way to the back of the store.

EL22 *Casework Lighting in Merchandise Sales Areas* (Climate Zones: all)

Casework lighting is not included in the tabulation of LPDs as long as it is integrated into the casework and is installed by the casework manufacturer. Follow recommendation EL2 for external casework accent lighting LPD above the base power allowance. Use accent lighting (fixture type C1 in Figure 5-23) to highlight key merchandise to "feature display" light levels (three to ten times the general merchandise). Lighting for casework must remain sensitive to the overall energy goals of the space. Strong consideration for internal display lighting should be given to energy-efficient or low-energy solutions, including linear fluorescent, fiber optic, and linear LED sources.

EL23 *Offices* (Climate Zones: all)

The target lighting in private offices is 30 average maintained footcandles for ambient lighting with a total of at least 50 footcandles provided on the desktop by a combination of the ambient and supplemental task lighting.

Private office plans account for approximately 5% of the floor area in retail buildings and should be limited to 0.95 W/ft^2 including circulation. The layout in Figure 5-24 is about 1.1 W/ft^2.

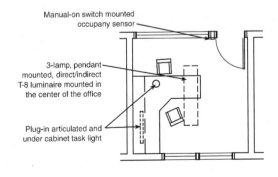

Figure 5-24. (EL23) Layout for office.

EL24   ***Warehousing/Active Storage Areas*** (**Climate Zones: all**)

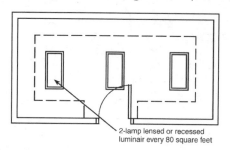

2-lamp lensed or recessed
luminair every 80 square feet

The target lighting in storage areas is 20–25 average maintained footcandles. Storage areas account for approximately 20% of the floor area and should be limited to 0.65 W/ft^2 including circulation. Use occupancy sensors or timers where appropriate.

Figure 5-25. (EL24) Layout for storage area.

EL25   ***Other Spaces*** (**Climate Zones: all**)

Lighting in the remaining 5% of the floor space is composed of various functions including restrooms, electrical/mechanical rooms, break rooms, workshops, and others. Limit the connected load in these spaces to 1.0 W/ft^2, which is equivalent to about one two-lamp high-performance T-8 luminaire every 64 ft^2. Use occupancy sensors or timers where appropriate.

References

NEEP. 2003. *Know-How Guide for Retail Lighting*. Lexington, MA: Northeast Energy Efficiency Partnerships. www.designlights.org/downloads/retail_guide.pdf.

IESNA. 1997. *EPRI Lighting Controls: Smart and Simple*. New York: Illuminating Engineering Society of North America.

IESNA. 2001. *ANSI/IESNA RP-2-01, Recommended Practice on Lighting Merchandising Areas (A Store Lighting Guide)*.

USGBC. 2005. LEED NC Indoor Environment Quality Credit 6.1, "Controllability of Systems: Lighting." Washington, DC: U.S. Green Building Council.

Exterior Lighting

Good Design Practice

Exterior lighting should be turned off or reduced by at least 50% one hour after normal business hours in response to light pollution and light trespass concerns. Maintain lighting that is required for safety and security.

EL26   ***Decorative Façade Lighting*** (**Climate Zones: all**)

Limit exterior decorative façade lighting to 0.2 W/ft^2 of illuminated surface. This does not include lighting of walkways or entry areas of the building that may also light the building itself. Façade lighting can provide additional attention to the retailer and improve feelings of safety and security. Limit the lighting equipment mounting locations to the building and do not install floodlights onto nearby parking lot lighting standards. Use downward-facing accent and sign lighting to comply with light trespass and light pollution concerns.

EL27   ***Sources*** (**Climate Zones: all**)

- All general lighting luminaires should utilize pulse-start metal halide, CMH, fluorescent, or compact fluorescent amalgam lamps with electronic ballasts.

- Standard high-pressure sodium lamps are not recommended due to their reduced visibility and poor color-rendering characteristics.

- Incandescent lamps are not recommended.

- For colder climates, fluorescent and CFL luminaires must be specified with cold-temperature ballasts. Use CFL amalgam lamps.

EL28   *Controls* **(Climate Zones: all)**

Use an astronomical time switch for all exterior lighting. Astronomical time switches are capable of retaining programming and time settings during loss of power for a period of at least ten hours. If a building energy management system is being used to control and monitor mechanical and electrical energy use, it can also be used to schedule and manage outdoor lighting energy use. Turn off exterior lighting not designated for security purposes when the building is unoccupied.

Parking lots and ground lighting are often beyond the control of the individual retailer and are not included here or in the Recommendation Tables in Chapter 3. Recommendations for parking lots and grounds are included in "Bonus Savings" sections EX1 through EX4 at the end of this chapter.

References

IESNA. 1998. *IESNA RP-20-1998, Recommended Practices and Design Guidelines.* New York: Illuminating Engineering Society of North America.

IESNA. 1999. *IESNA RP-33-99, Recommended Practices and Design Guidelines.* New York: Illuminating Engineering Society of North America.

IESNA. 1994. *IESNA DG-5-94, Recommended Practices and Design Guidelines.* New York: Illuminating Engineering Society of North America.

IESNA. 2003. *IESNA G-1-03, Recommended Practices and Design Guidelines.* New York: Illuminating Engineering Society of North America.

LRC. 1996. *Outdoor Lighting Pattern Book.* Troy, NY: Lighting Research Center.

USGBC. 2005. LEED NC Sustainable Sites Credit 8, "Light Pollution Reduction." Washington, DC: U.S. Green Building Council.

USGBC. 2005. LEED NC Indoor Environment Quality Credit 6.1, "Controllability of Systems: Lighting." Washington, DC: U.S. Green Building Council.

HVAC

Good Design Practice

HV1   *General* **(Climate Zones: all)**

The HVAC equipment for this Guide includes packaged-unit systems and split systems generally referred to as *air-conditioning* or *heat pump* units (see Figure 5-26). These systems are suitable for projects with no central plant. This Guide does not cover water-source or ground-source heat pumps, systems that use liquid water chillers or purchased chilled water for cooling, or oil, hot water, solar, steam, or purchased steam for heating. These systems are alternative means that may be used to achieve the energy savings target of this Guide.

The systems included in this Guide are available in pre-established increments of capacity with a refrigeration cycle and heating source. The components are factory designed and assembled and include fans, motors, filters, heating source, cooling coil, refrigerant compressor, controls, and condenser. The components can be in a single package or a split system that separates the evaporator and condenser sections.

Performance characteristics vary among manufacturers, and the selected equipment should match the calculated heating and cooling loads (sensible and latent), also taking into account the importance of meeting latent cooling loads under part-load conditions. See HV3 for calculating the loads, HV4 for meeting latent cooling loads under part-load conditions, and HV13 for recommendations on zoning the building. See HV21 for a discussion on the location of space thermostats. The equipment should be listed as being in conformance with electrical and safety standards with its performance ratings certified by a nationally recognized certification program.

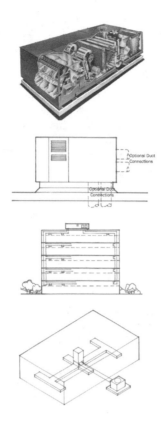

Single packaged units can be mounted on the roof, at grade level, or indoors. Split systems generally have the blower unit, including filters and coils, located indoors or in unconditioned space and the condensing unit outdoors on the roof or at grade level. On smaller systems, the blower is commonly incorporated in an indoor furnace section. The blower unit may also be located outdoors; if so, it should be mounted on the roof to avoid ductwork outside the building envelope. See HV9 for further discussion on the ductwork recommendations. The equipment should be located in a position that results in minimizing fan power, ducting, and wiring. See Figure 5-26 for examples of typical HVAC equipment and duct system layouts.

Figure 5-26. (HV1) Typical HVAC equipment and duct system layouts.

HV2   HVAC System Types (Climate Zones: all)

This Guide considers packaged-unit systems and split systems with a refrigerant-based direct expansion system for electric cooling and heating by means of one of the three following options:

> Option 1: Indirect gas-fired heater
> Option 2: Electric resistance heat
> Option 3: Heat pump unit

Indirect gas-fired heaters use a heat exchanger as part of the factory-assembled unit to separate the burner and products of combustion from the circulated air.

Electric resistance heaters can be part of the factory-assembled unit or can be installed in the duct distribution system.

The auxiliary heat source for heat pump units may also be used to supply heating to the space during the defrost cycle and can be either electric or gas.

Where variable air volume systems are used, the refrigeration system requires reduced capacity in response to reduced load. The package unit should be designed to maintain the required apparatus dew point for humidity control. The controls of a variable air volume system should be arranged to reduce the supply air to the minimum setpoint for ventilation before tempering of the air occurs. Variable-speed drives should be considered as an option to reduce airflow and fan/motor energy.

HV3   ***Cooling and Heating Loads* (Climate Zones: all)**

Cooling and heating system design loads should be calculated in accordance with generally accepted engineering standards and handbooks such as *ASHRAE Handbook— Fundamentals* or *ACCA Manual N*. Any safety factor should be applied cautiously and only to a building's internal loads to prevent oversizing of equipment. If the unit is over-sized and the cooling capacity reduction is limited, short-cycling of compressors could occur and the system may not have the ability to dehumidify the building properly. Include the cooling and heating load of the outdoor air to determine the total cooling and heating requirements of the unit. In determining cooling requirements, the sensible and latent load to cool the outdoor air to room temperature must be added to the build-ing cooling load. For heating, the outdoor air brought into the space must be heated to the room temperature and the heat required added to the building heat loss. On variable air volume systems, the minimum supply airflow to a zone should comply with local code, the current ASHRAE Standard 62, and the current ASHRAE Standard 90.1 and should be taken into account in calculating heating loads of the outdoor air.

HV4   ***Humidity Control* (Climate Zones: all)**

The sensible load in the building does not decrease proportionately with the latent load, and as a result, the space relative humidity will tend to increase under cooling part-load conditions. Select systems with cooling part-load performance that minimizes the number of hours that the space relative humidity remains above 60%. For part loads and variable air volume systems, multiple compressors are desirable to reduce the capacity as low as possible to meet the minimum cooling requirements and operate efficiently at part loads. On systems with multiple compressors, the compressors turn on or off or unload to maintain the space air temperature setpoint. On systems that employ supply air temperature reset, controls must be added to ensure that the relative humidity within the space does not exceed 60%.

HV5   ***Energy Recovery* (Climate Zones: all)**

Total energy recovery equipment can provide an energy-efficient means of dealing with the latent and sensible outdoor air cooling loads during peak summer conditions. It can also reduce the required heating of outdoor air in cold climates.

Exhaust air energy recovery can be provided through a separate energy recovery ventilator (ERV) that conditions the outdoor air before entering the air-conditioning or heat pump unit, an ERV that attaches to an air-conditioning or heat pump unit, or an air-conditioning or heat pump unit with the ERV built into it.

For maximum benefit, energy recovery designs should provide as close to balanced outdoor and exhaust airflows as is practical, taking into account the need for building pressurization and any exhaust that cannot be incorporated into the system.

Exhaust for ERVs may be taken from spaces requiring exhaust (using a central exhaust duct system for each unit) or directly from the return airstream (as with a uni-tary accessory or integrated unit).

Where economizers are used with an ERV, the energy recovery system should be controlled in conjunction with the economizer and provide for the economizer function. Where energy recovery is used without an economizer, the energy recovery system should be controlled to prevent unwanted heat and an outdoor air bypass of the energy recovery equipment should be used. In cold climates, manufacturer's recommendations for frost control should be followed.

HV6   *Equipment Efficiency* **(Climate Zones: all)**

The cooling equipment should meet or exceed the listed seasonal energy efficiency ratio (SEER) or energy efficiency ratio (EER) for the required capacity. The cooling equipment should also meet or exceed the integrated part-load value (IPLV) where shown.

Heating equipment should meet or exceed the listed annual fuel utilization efficiency (AFUE) or thermal efficiency for indirect gas-fired heaters at the required capacity. For heat pump applications, the heating efficiency should meet or exceed the listed heating seasonal performance factor (HSPF) or coefficient of performance (COP) for the required capacity based on 47°F outdoor air temperature.

HV7   *Ventilation Air* **(Climate Zones: all)**

The amount of outdoor air should be based on ASHRAE Standard 62.1-2004 but in no case be less than the values required by local code. The number of people used in computing the ventilation quantity required should be based on either known occupancy, local code, or Standard 62.1-2004. For retail sales, Standard 62.1-2004 suggests 16 cfm per person based on 15 people per 1,000 ft^2. Ventilation guidelines for pharmacies, photo processing areas, receiving docks, and warehousing spaces are detailed in the standard.

Each air-conditioning or heat pump system should have an outdoor air connection through which ventilation air is introduced and mixes with the return air. The outdoor air can be mixed with the return air either in the ductwork prior to the air-conditioning or heat pump unit or at the unit's mixing plenum. In either case, the damper and duct/ plenum should be arranged to promote mixing and minimize stratification. A typical design for the ventilation system of a small retail building is shown in Figure 5-27.

An air economizer mode can save energy by using outdoor air for cooling in lieu of mechanical cooling when the temperature of the outdoor air is low enough to meet the cooling needs. The system should be capable of modulating the outdoor air, return air, and relief air dampers to provide up to 100% of the design supply air quantity as outdoor air for cooling.

Systems should use a motorized outdoor air damper instead of a gravity damper to prevent outdoor air from entering during the unoccupied periods when the unit may recirculate air to maintain setback or setup temperatures. The motorized outdoor air damper for all climate zones should be closed during the entire unoccupied period except when it may open in conjunction with an economizer cycle.

Demand control ventilation should be used in areas that have varying and high occupancy loads during the occupied periods to vary the amount of outdoor air in response to the need in a zone. The amount of outdoor air could be controlled by carbon

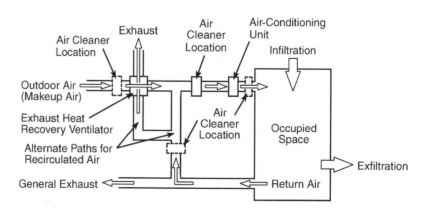

Figure 5-27. (HV7) Example of ventilation system.

dioxide sensors, as a proxy for the adequacy of ventilation, that measure the change in carbon dioxide levels in a zone relative to the levels in the outdoor air. A controller will operate the outdoor air, return air, and relief air dampers to maintain proper ventilation. See HV22 for more discussion on demand ventilation control and HV14 for methods of operating the system efficiently.

HV8   *Exhaust Air* **(Climate Zones: all)**

Central exhaust systems for restrooms, janitorial closets, etc., should be interlocked to operate with the air-conditioning or heat pump unit except during unoccupied periods. These exhaust systems should have a motorized damper that opens and closes with the operation of the fan. The damper should be located as close as possible to the duct penetration of the building envelope to minimize conductive heat transfer through the duct wall and avoid having to insulate the duct. During unoccupied periods, the damper should remain closed and the exhaust fan turned off, even while the air-conditioning or heat pump unit is operating, to maintain setback or setup temperatures.

HV9   *Ductwork Distribution* **(Climate Zones: all)**

Many retail stores use unitary rooftop systems with no ductwork or minimal ductwork. This is a less desirable design option that should be avoided because nonducted systems have more difficulty achieving proper airflow and ventilation, often resulting in poor air quality.

Air should be ducted through low-pressure ductwork with a system pressure classification of less than 2 in. Where practical, rigid ductwork is preferred. Supply air should be ducted to supply diffusers in each individual space. Return air should be ducted from return registers provided in appropriate locations for proper airflow but not necessarily in every space. The ductwork should be as direct as possible, minimizing the number of elbows, abrupt contractions and expansions, and transitions. Long-radius elbows and 45° lateral take-offs should be used wherever possible. Where variable air volume systems are used, they should have single-duct air terminal units to control the volume of air to the zone based on the space temperature sensor.

In general, the following sizing criteria should be used for the duct system components:

a. Diffusers and registers should be sized with a static pressure drop no greater than 0.08 in.
b. Supply and return ductwork should be sized with a pressure drop no greater than 0.08 in. per 100 linear feet of duct run.
c. Flexible ductwork should be of the insulated type and should be

1. limited to connections between duct branch and diffusers,
2. limited to connections between duct branch and variable air volume terminal units,
3. limited to 5 ft (fully stretched length) or less,
4. installed without any kinks,
5. installed with a durable elbow support when used as an elbow, and
6. installed with no more that 15% compression from fully stretched length, and
7. hanging straps, if used, need to use a saddle to avoid crimping the inside cross-sectional area. For ducts with 12 in. or less diameter use a 3 saddle; for larger than 12 in. use a 5 in. saddle.

Ductwork should not be installed outside the building envelope in order to minimize heat gain to or heat loss from the ductwork due to outdoor air temperatures and solar heat gain. Ductwork on rooftop units should enter or leave the air-conditioning or heat pump unit through an insulated roof curb around the perimeter of the air-conditioning or heat pump unit's footprint.

Duct board should be airtight (duct seal level B, from ASHRAE Standard 90.1) and should be taped and sealed with products that maintain adhesion. Duct static pressures should be designed and equipment and diffuser selections should be selected to not exceed the noise criteria for the space. See HV19 for additional information.

HV10   Duct Insulation (Climate Zones: all)

All supply air ductwork should be insulated. All return air ductwork located above the ceiling and below the roof should be insulated. Any outdoor air ductwork should be insulated. All exhaust and relief air ductwork between the motor-operated damper and penetration of the building exterior should be insulated.

Include a vapor barrier on the outside of the insulation where condensation is possible.

Exception: In conditioned spaces without a finished ceiling, only the supply air duct mains and major branches should be insulated. Individual branches and run-outs to diffusers in the space being served do not need to be insulated, except where it may be necessary to prevent condensation.

HV11   Duct Sealing and Leakage (Climate Zones: all)

The ductwork should be sealed for Seal Class B from ASHRAE Standard 90.1 and leak-tested at the rated pressure. The leakage should not exceed the allowable cfm/100 ft^2 of duct area for the seal and leakage class of the system's air quantity apportioned to each section tested. See HV15 for guidance on ensuring the air system performance.

HV12   Fan Motors (Climate Zones: all)

Motors for fans 1 hp or greater should meet National Electrical Manufacturers Association (NEMA) premium efficiency motor guidelines when available.

HV13   Thermal Zones (Climate Zones: all)

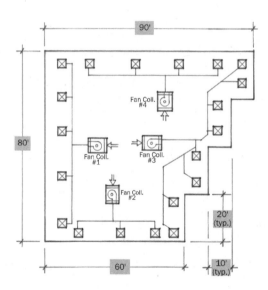

Figure 5-28. (HV13) Perimeter system zoning.

Retail buildings should be divided into thermal zones based on building size, part-load performance requirements, space layout and function, number of tenants, and the needs of the user. In a retail building with similar internal loads throughout, a minimum of one zone for the interior and one for the front exposure is recommended. If side exposures have significant glass area, additional zones may be needed. (See Figure 5-28 for an example of perimeter system zoning.)

Zoning can also be accomplished using multiple air-handling units or by having multiple-zone control with a single air-handling unit. The temperature sensor for a zone should be installed in a location representative of that entire zone.

HV14   ***Control Strategies*** **(Climate Zones: all)**

Control strategies can be designed to help reduce energy. Time-of-day scheduling is useful when it is known which portions of the building will have reduced occupancy. Control of the ventilation air system can be tied into this control strategy.

Having a setback temperature for unoccupied periods during the heating season or setup temperature during the cooling season will help to save energy. A pre-occupancy operation period will help to purge the building of contaminants that build up overnight from the outgassing of products and packaging materials. In hot, humid climates, care should be taken to avoid excessive relative humidity conditions during unoccupied periods.

HV15   ***Testing, Adjusting, and Balancing (TAB)*** **(Climate Zones: all)**

After the system has been installed, cleaned, and placed in operation, the system should be tested, adjusted, and balanced for proper operation. This procedure will help to ensure that the correctly sized diffusers, registers, and grilles have been installed, that each space receives the required airflow, that the equipment meets the intended performance, and that the controls operate as intended. The TAB subcontractor should certify that the instruments used in the measurement have been calibrated within 12 months prior to use. A written report should be submitted for inclusion in the O&M manuals.

HV16   ***Filters*** **(Climate Zones: all)**

Air-conditioning and heat pump unit filters are included as part of the factory-assembled unit and should be at least MERV 8, based on ASHRAE Standard 52.2. Use a filter differential pressure gauge to monitor the pressure drop across the filters. The gauge should be checked and the filter should be inspected on a routine basis. Filters should be replaced when their pressure drop exceeds the filter manufacturer's recommendations for replacement or when visual inspection indicates the need. A monitor should be included to send an alarm if the predetermined pressure drop is exceeded. Upon completion of construction, all filters should be replaced prior to building occupancy.

Cautions

HV17   ***Heating Sources*** **(Climate Zones: all)**

Forced-air electric resistance and gas-fired heaters require a minimum airflow rate to operate safely. These systems, whether stand-alone or incorporated into an air-conditioning or heat pump unit, should include factory-installed controls to shut down the heater when there is inadequate airflow resulting in high air temperatures.

HV18   ***Return and Relief Air*** **(Climate Zones: all)**

Relief (rather than return) fans or blowers should be used when necessary to maintain building pressurization during economizer operation. However, where return duct static pressure exceeds 0.5 in. of water, return fans should be used.

HV19 **Noise Control (Climate Zones: all)**

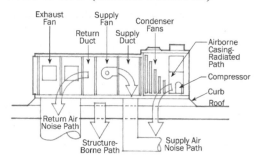

Figure 5-29. (HV19) Typical noise paths for roof-top-mounted HVAC units.

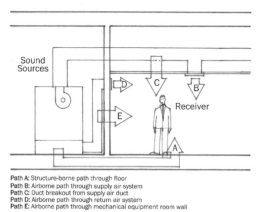

Path A: Structure-borne path through floor
Path B: Airborne path through supply air system
Path C: Duct breakout from supply air duct
Path D: Airborne path through return air system
Path E: Airborne path through mechanical equipment room wall

Figure 5-30. (HV19) Typical noise paths for interior-mounted HVAC units.

Acoustical requirements may necessitate attenuation of the noise associated with the supply and/or return air, but the impact on fan energy consumption should also be considered and, if possible, compensated for in other duct or fan components. Acoustical concerns may be particularly critical in short, direct runs of ductwork between the fan and supply or return outlet.

Where practical, avoid installation of the air-conditioning or heat pump units above areas that customers visit. Consider locations above less critical spaces such as storage areas, restrooms, corridors, etc. (See Figures 5-29 and 5-30 for typical noise paths for HVAC units.)

HV20 **Heating Supply Air Temperatures (Climate Zones: all)**

Ducts and supply air registers should be selected based on discharge air temperature and flow rate.

HV21 **Zone Temperature Control (Climate Zones: all)**

The number of spaces in a zone and the location of the temperature-sensing point will affect the control of temperature in the various spaces of a zone. Locating the thermostat in one room of a zone with multiple spaces provides feedback based only on the conditions of that room. Locating a single thermostat in a large open area may provide a better response to the conditions of the zone with multiple spaces. Selecting the room or space that will best represent the thermal characteristics of the space due to both external and internal loads will provide the greatest comfort level.

To prevent misreading of the space temperature, zone thermostats should not be mounted on an exterior wall. Where this is unavoidable, use an insulated sub-base for the thermostat.

HV22 ***Carbon Dioxide Sensors* (Climate Zones: all)**

The number and location of carbon dioxide sensors for demand-control ventilation can affect the representation of the building or zone occupancy. A minimum of one CO_2 sensor per zone is recommended for systems with greater than 500 cfm of outdoor air. Multiple sensors may be necessary if the ventilation system serves spaces with significantly different occupancy expectations. Where multiple sensors are used, the ventilation should be based on the sensor recording the highest concentration of CO_2.

Sensors used in merchandise areas with high outdoor air requirements should be installed in the return air ducts to provide an average CO_2 measurement for the zone. For sensors mounted in return air duct, adequate access for sensor calibration and field testing must be provided. The number and location of sensors should take into account the sensor manufacturer's recommendations for the particular product.

The demand ventilation controls should maintain CO_2 concentrations less than or equal to 600 ppm plus the outdoor air CO_2 concentration in all spaces with CO_2 sensors. However, the outdoor air ventilation rate should not exceed the maximum design outdoor air ventilation rate required by code regardless of CO_2 concentration.

The outdoor air CO_2 concentration can be assumed to be 400 ppm without any direct measurement, or the CO_2 concentration can be monitored using a CO_2 sensor located near the position of the outdoor air intake.

CO_2 sensors should be certified by the manufacturer to have an error of 75 ppm or less and be factory calibrated. Inaccurate CO_2 sensors can cause excessive energy consumption and poor air quality, so these need to be calibrated as recommended by the manufacturer.

HV23 ***Economizers* (Climate Zones: ❸ ⑤ ⑥ ❼ ❽)**

Economizers, when recommended, should be employed on air conditioners to help save energy by providing free cooling when ambient conditions are suitable to meet all or part of the space cooling load. Consider using enthalpy controls (versus dry-bulb temperature controls) to help ensure that unwanted moisture is not introduced into the space in hot, humid climates. Economizers are not recommended in climate zone 1. There may be some applicability in dry climate areas in climate zone 2. Periodic maintenance is important with economizers, as dysfunctional economizers can cause substantial excess energy consumption due to malfunctioning dampers and/or sensors.

References

ASHRAE Handbook—HVAC Applications. Atlanta: American Society of Heating, Refrigerating and Air-Conditioning Engineers, Inc.

ASHRAE Handbook—Fundamentals. Atlanta: American Society of Heating, Refrigerating and Air-Conditioning Engineers, Inc.

ASHRAE Handbook—HVAC Systems and Equipment. Atlanta: American Society of Heating, Refrigerating and Air-Conditioning Engineers, Inc.

ASHRAE. 2004. *ASHRAE Standard 62.1-2004 - Ventilation for Acceptable Indoor Air Quality.* Atlanta: American Society of Heating, Refrigerating and Air-Conditioning Engineers, Inc.

National Electrical Manufacturers Association, www.nema.org, Standards and Publications section.

SERVICE WATER HEATING

**Good Design
Practice**

WH1   *Service Water Heating Types* **(Climate Zones: all)**

The service water heating (SWH) equipment discussed in this Guide is considered to use the same type of fuel source used for the HVAC system. This Guide does not cover systems that use oil, hot water, steam, or purchased steam for generating SWH. The Guide also does not address the use of solar or site-recovered energy (including heat pump water heaters). These systems are alternative means that may be used to achieve 30% or greater savings over ASHRAE Standard 90.1-1999 and, where used, the basic principles of this Guide would apply.

The SWH equipment included in this Guide for the HVAC options listed in HV2 are the gas-fired water heater and the electric water heater.

Both natural gas and propane fuel sources are available options for gas-fired units.

WH2   *System Description* **(Climate Zones: all)**

1. **Gas-fired storage water heater.** A water heater with a vertical or horizontal water storage tank. A thermostat controls the delivery of gas to the heater's burner. The heater requires a vent to exhaust the products of combustion.
2. **Gas-fired instantaneous water heater.** A water heater with minimal water storage capacity. The heater requires a vent to exhaust the products of combustion. An electronic ignition is recommended to avoid the energy losses from a standing pilot.
3. **Electric resistance storage water heater.** Water heater consisting of a vertical or horizontal storage tank with one or more immersion heating elements. Thermostats controlling heating elements may be of the immersion or surface-mounted type. For typical retail applications, small water heaters are available from 2 to 20 gal.
4. **Electric resistance instantaneous water heater.** Compact, under-cabinet, or wall-mounted types with insulated enclosure and minimal water storage capacity; a thermostat controls the heating element, which may be of the immersion or surface-mounted type. Instantaneous, point-of-use water heaters should provide water at a constant temperature regardless of input water temperature.

WH3   *Sizing* **(Climate Zones: all)**

The water heating system should be sized to meet the anticipated peak hot water load, typically 0.4 gal per hour per store employee in the average retail building. The hot water demand will be higher if showers or other high-volume uses exist, and these should be accounted for in sizing equipment. The supply water temperature should be no higher than 120°F to avoid injuries due to scalding.

WH4   *Equipment Efficiency* **(Climate Zones: all)**

Efficiency levels are provided in the Guide for gas instantaneous, gas-fired storage, and electric resistance storage water heaters. For gas-fired instantaneous water heaters, the energy factor and thermal efficiency levels correspond to commonly available instantaneous water heaters.

The gas-fired storage water heater efficiency levels correspond to condensing storage water heaters. High-efficiency, condensing gas storage water heaters (energy factor > 0.90 or thermal efficiency > 0.90) are alternatives to the use of gas-fired instantaneous water heaters.

Electric storage water heater efficiency should be calculated as 0.99 − 0.0012 × water heater volume. Instantaneous electric water heaters are an acceptable alternative to high-efficiency storage water heaters. Electric instantaneous water heaters are more efficient than electric storage water heaters, and point-of-use versions will minimize piping losses. However, their impact on building peak electric demand can be significant and should be taken into account during design.

WH5   *Location* **(Climate Zones: all)**

The water heater should be located close to the hot water fixtures to avoid the use of a hot water return loop or the use of heat tracing on the hot water supply piping. Where electric resistance heaters are used, point-of-use water heaters should be considered when there is a low number of fixtures or where they can eliminate the need for a recirculating loop.

WH6   *Pipe Insulation* **(Climate Zones: all)**

All SWH piping should be installed in accordance with accepted industry standards. Insulation levels should be in accordance with the recommendation levels in Chapter 3, and the insulation should be protected from damage. Include a vapor retardant on the outside of the insulation.

Reference   *ASHRAE Handbook—HVAC Applications.* Atlanta: American Society of Heating, Refrigerating and Air-Conditioning Engineers, Inc.

BONUS SAVINGS

Plug Loads

Building owners and other users of this Guide can benefit from additional energy savings by outfitting retail stores with efficient appliances, office equipment, and other devices plugged into electric outlets. These "plug loads" can account for 4% to 40% of a retail building's annual energy requirements and energy expense.[1] In addition to their own energy requirements, plug loads also are a source of internal heat gains that increase air-conditioning energy use. There is a large variation of these loads depending on the retail operation and the need for retail displays, communications equipment, computing requirements, product movement, and cleaning requirements.

Retail displays may have specialized task lighting, ventilation fans for cooling, and refrigerated end-caps (cases that sell soda and other refrigerated items), just to name a few. Most retail operations have small computer networks that may include inventory systems and point-of-sale devices. In many retail operations, Internet terminals provide customers the ability to order products online. Retail stores also have offices for managers and staff; these spaces include devices such as fax machines, calculators, and copiers. Employee break rooms are often equipped with refrigerators, microwave ovens, and coffee makers. Many retail stores have vending machines that may or may not be publicly accessible. Cleaning equipment, such as waxers and buffers, can be directly connected or recharged. Telephone switches, fire and security alarm systems, and energy management systems all contribute to the building loads. Design teams should identify special loads in retail spaces (examples might include lighting displays in hardware stores and fish tanks in pet stores) and try to reduce these loads as appropriate. The team should take inventory in an existing similar facility to fully understand the loads and determine possible points for energy savings.

The recommendations presented in Table 5-2 for purchase and operation of plug load equipment are an integral part of this Guide, but the energy savings from the plug load recommendations are expected to be in addition to the target 30% savings.

---

1. Data available at www.eia.doe.gov/emeu/cbecs/cbec-eu4.pdf.

Table 5-2. Recommendations for Efficient Plug Load Equipment

Equipment/Appliance Type	Purchase Recommendation	Operating Recommendation
Desktop computer	ENERGY STAR® only	Implement sleep mode software
Electronic cash registers and point-of-sale devices	Purchase with flat-screens with sleep modes	Many of these items are only used during peak times and excess equipment should be turned off
Laptop computer—use where practical instead of desktops to minimize energy use	ENERGY STAR® only	Implement sleep mode software
Computer monitors (may include point-of-sale monitors)	ENERGY STAR® flat-screen monitors only	Implement sleep mode software
Printer	ENERGY STAR® only	Implement sleep mode software
Copy machine	ENERGY STAR® only	Implement sleep mode software
Fax machine	ENERGY STAR® only	Implement sleep mode software
Water cooler	ENERGY STAR® only	N/A
Refrigerator	ENERGY STAR® only	N/A
Vending machines	ENERGY STAR® only	Delamp display lighting

PL1   Purchase Energy-Efficient Equipment (Climate Zones: all)

Many of the plug load items in retail stores can be specified to be energy efficient. ENERGY STAR® labels are applicable for computer equipment, office equipment, and vending machines. Refrigerators for break rooms should also be ENERGY STAR® rated. To further reduce energy use, make sure that the ENERGY STAR® sleep modes are enabled on computer and copier equipment. The use of such equipment in most retail operations is limited and short delays to sleep mode are appropriate.

PL2   Use Time Clocks to Disable Plug Loads (Climate Zones: all)

Time clocks can be used to effectively disable loads when these loads are not needed. Display lighting end-cap displays can be de-energized when the store is closed. This may provide a narrower window than the occupied hours of the store because of stocking times. It may be possible to use common electrical circuits on central time clocks or the energy management system. This technique can also be used to turn off vending machines, coffee makers, point-of-sale devices, calculators, credit card dialers, and other miscellaneous loads when the building is not occupied. Care must be exercised not to turn on all the loads when the HVAC and lighting are energized to avoid a peak demand on start-up.

PL3   Use Motion-Based Plug Strips (Climate Zones: all)

Plug strips with motion sensors can be deployed in break rooms and offices to turn off plug loads such as vending machines, computer monitors, calculators, and other equipment that plugs in (usually identified with power packs or cubes). Motion sensor circuits can also be used for end-cap displays and lit displays. In some cases it is more dramatic to have items "turn on" as customers approach to gain attention.

PL4   ***Delay Loads to Off-Peak Hours, if Possible* (Climate Zones: all)**

Delaying loads to off-peak hours probably will not save energy but may save on utility costs. Examples might be electric pallet jacks, forklifts, and chargeable cleaning equipment. Most of these items are used and plugged in near the end of daily operation, causing peak loads coincident with lighting loads. Time clocks can be put on these circuits to delay operation to times when the store is not occupied. Other examples might include dishwashers in kitchen areas and washing machines, available to staff in some cases.

PL5   ***Identify Loads That Are Not Needed* (Climate Zones: all)**

Some loads can be disconnected. Quite often in cooler climates, water coolers run refrigerated units when the temperature of the street water is adequate. Excess equipment should be disconnected. Point-of-sale equipment should only be energized when it will be needed. In many cases, some of the point-of-sale equipment is only used during the Christmas season. Vending machines can be delamped in non-public areas.

Reference   DOE. ENERGY STAR®. www.energystar.gov.

Exterior Lighting

Good Design Practice

The following recommendations are not included in the Recommendation Tables in Chapter 3 because parking lots and grounds are often beyond the control of the individual retailer. If designing for parking lots and grounds, follow recommendations EX1 through EX4.

EX1   ***Exterior Lighting Power* (Climate Zones: all)**

Limit exterior lighting power to 0.15 W/ft^2 for parking lot and grounds lighting. Calculate only for paved areas, excluding grounds that do not require lighting.

EX2   ***Sources* (Climate Zones: all)**

- All general lighting luminaires should utilize pulse-start metal halide, fluorescent, induction, or compact fluorescent amalgam lamps with electronic ballasts.
- Standard high-pressure sodium lamps are not recommended due to their reduced visibility and poor color-rendering characteristics.
- Incandescent lamps are not recommended.
- For colder climates, fluorescent and compact fluorescent lamp (CFL) luminaires must be specified with cold-temperature ballasts. Use CFL amalgam lamps.

EX3   ***Parking Lighting* (Climate Zones: all)**

Parking lot lighting locations should be coordinated with landscape plantings so that tree growth does not block effective lighting from pole-mounted luminaires.

Parking lot lighting should not be significantly brighter than lighting of the adjacent street. Follow IESNA RP-33-1999 recommendations for uniformity and illuminance recommendations.

For parking lot and grounds lighting, do not increase luminaire wattage in order to use fewer lights and poles. Increased contrast makes it harder to see at night beyond the immediate fixture location. Flood lights and non-cutoff wall-packs should not be used, as they cause hazardous glare and unwanted light encroachment on neighboring properties. Limit lighting in parking and drive areas to not more than 360-watt pulse-start metal halide lamps at a maximum 25 ft mounting height in urban and suburban areas.

Use cutoff luminaries that provide all light below the horizontal plane and help eliminate light trespass.

The use of cutoff luminaires and limiting overall site brightness also permits greater visibility of storefronts from more distant locations of the parking areas and adjacent roadways, permitting retailers greater opportunity to create off-site visual impact.

EX4 ***Controls* (Climate Zones: all)**

Use an astronomical time switch for all exterior lighting. Astronomical time switches are capable of retaining programming and the time setting during loss of power for a period of at least 10 hours. If a building energy management system is being used to control and monitor mechanical and electrical energy use, it can also be used to schedule and manage outdoor lighting energy use. Turn off exterior lighting not designated for security purposes when the building is unoccupied.

References

IESNA. 1998. *IESNA RP-20-1998, Recommended Practices and Design Guidelines.* New York: Illuminating Engineering Society of North America.

IESNA. 1999. *IESNA RP-33-99, Recommended Practices and Design Guidelines.* New York: Illuminating Engineering Society of North America.

IESNA. 1994. *IESNA DG-5-94, Recommended Practices and Design Guidelines.* New York: Illuminating Engineering Society of North America.

IESNA. 2003. *IESNA G-1-03, Recommended Practices and Design Guidelines.* New York: Illuminating Engineering Society of North America.

LRC. 1996. *Outdoor Lighting Pattern Book.* Troy, NY: Lighting Research Center.

Appendix A
Envelope Thermal Performance Factors

The Recommendation Tables in Chapter 3 present the opaque envelope recommendations in a standard format. This is a simple approach, but it limits the construction options. In order to allow for alternative constructions, the recommendations can also be represented by thermal performance factors such as U-factors for above-grade components, C-factors for below-grade walls, or F-factors for slabs-on-grade; see Table A-1. Any alternative construction that is less than or equal to these thermal performance factors will be acceptable alternatives to the recommendations.

Table A-1. Envelope Thermal Performance Factors

Item	Description	Unit	#1	#2	#3	#4	#5	#6
Roof	Insulation entirely above deck	R	15	20	25	30		
		U	0.063	0.048	0.039	0.032		
	Metal building	R	19	13+13	13+19	16 + 19		
		U	0.065	0.055	0.049	0.047		
	Attic and other	R	30	38	49	60		
		U	0.034	0.027	0.021	0.017		
	Single rafter	R	30	38	38+5	38+10		
		U	0.036	0.028	0.024	0.022		
Walls, Above Grade	Mass	R	5.7	7.6	9.5	11.4	13.3	15.2
		U	0.151	0.123	0.104	0.090	0.080	0.071
	Metal building	R	13	13+13	13+16			
		U	0.113	0.057	0.055			
	Steel-framed	R	13	13+3.8	13+7.5	13+10		
		U	0.124	0.084	0.064	0.055		
	Wood-framed and other	R	13	13+3.8	13+7.5	13+10		
		U	0.089	0.064	0.051	0.045		
Walls, Below Grade	Below-grade	R	7.5	12.5	15			
		C	0.119	0.075	0.063			
Floors	Mass	R	4.2	6.3	10.4	12.5	14.6	16.7
		U	0.137	0.107	0.074	0.064	0.056	0.051
	Steel joist	R	19	30	38			
		U	0.052	0.038	0.032			
	Wood-framed and other	R	19	30				
		U	0.051	0.033				
Slabs	Unheated	R-in.	10-24	15-24	20-24			
		F	0.54	0.52	0.51			
	Heated	R-in.	7.5-12	7.5-24	10-36	10 Full	15 Full	20 Full
		F	1.02	0.95	0.84	0.55	0.44	0.373

Appendix B
Additional Resources

BOOKS AND STANDARDS

AAMA. 2003. *AAMA 507, Standard Practice for Determining the Thermal Performance Characteristics of Fenestration Systems Installed in Commercial Buildings*. Schaumburg, IL: American Architectural Manufacturers Association.

ASHRAE. 1999. *ANSI/ASHRAE/IESNA Standard 90.1-1999, Energy Standard for Buildings Except Low-Rise Residential Buildings*. Atlanta: American Society of Heating, Refrigerating and Air-Conditioning Engineers, Inc.

ASHRAE. 2001. *ANSI/ASHRAE/IESNA 90.1-2001, Energy Standard for Buildings Except Low-Rise Residential Buildings*. Atlanta: American Society of Heating, Refrigerating and Air-Conditioning Engineers, Inc.

ASHRAE. 2004. *Advanced Energy Design Guide for Small Office Buildings*. Atlanta: American Society of Heating, Refrigerating and Air-Conditioning Engineers, Inc.

ASHRAE. 2004. *ANSI/ASHRAE Standard 62.1-2004, Ventilation for Acceptable Indoor Air Quality*. Atlanta: American Society of Heating, Refrigerating and Air-Conditioning Engineers.

ASHRAE. 2004. *ANSI/ASHRAE/IESNA Standard 90.1-2004, Energy Standard for Buildings Except Low-Rise Residential Buildings*. Atlanta: American Society of Heating, Refrigerating and Air-Conditioning Engineers, Inc.

ASHRAE. 2005. *ASHRAE Handbook—Fundamentals*. Atlanta: American Society of Heating, Refrigerating and Air-Conditioning Engineers, Inc.

ASHRAE. 2006. *ANSI/ASHRAE Standard 169, Weather Data for Building Design Standards*. Atlanta: American Society of Heating, Refrigerating and Air-Conditioning Engineers, Inc.

ASTM. 2001. *ASTM E 1980, Standard Practice for Calculating Solar Reflectance Index of Horizontal and Low-Sloped Opaque Surfaces*. West Conshohocken, PA: American Society for Testing and Materials.

ASTM. 2003. *ASTM E 2178, Standard Test Method for Air Permeance of Building Materials*. West Conshohocken, PA: American Society for Testing and Materials.

Evans, Benjamin. 1997. *Daylighting Design, Time Saver Standards for Architectural Design Data.* New York: McGraw-Hill.

IESNA. 1994. *IESNA DG-5-94, Recommended Practices and Design Guidelines.* New York: Illuminating Engineering Society of North America.

IESNA. 1996. *EPRI Lighting Controls—Patterns for Design.* New York: Illuminating Engineering Society of North America.

IESNA. 1997. *EPRI Daylight Design: Smart and Simple.* New York: Illuminating Engineering Society of North America.

IESNA. 1998. *IESNA RP-20-1998, Recommended Practices and Design Guidelines.* New York: Illuminating Engineering Society of North America.

IESNA. 1999. *IESNA RP-33-99, Recommended Practices and Design Guidelines.* New York: Illuminating Engineering Society of North America.

IESNA. 2001. *ANSI/IESNA RP-2-01, Recommended Practice on Lighting Merchandising Areas (A Store Lighting Guide).*

IESNA. 2003. *IESNA G-1-03, Recommended Practices and Design Guidelines.* New York: Illuminating Engineering Society of North America.

LBNL. 1997. Tips for daylighting with windows. *Windows & Daylighting.* Berkeley, CA: Lawrence Berkeley National Laboratories. http://windows.lbl.gov/daylighting/designguide/designguide.html.

LRC. 1996. *Outdoor Lighting Pattern Book.* Troy, NY: Lighting Research Center. Available for purchase from www.lightingresearch.org.

NBI. 2003. *Advanced Lighting Guidelines.* White Salmon, WA: New Buildings Institute. www.newbuildings.org/lighting.htm.

NEEP. 2003. *Know-How Guide for Retail Lighting.* Lexington, MA: Northeast Energy Efficiency Partnerships. Available as free download from www.designlights.org.

Torcellini, P.A., D.B. Crawley. 2006. Understanding zero-energy buildings. *ASHRAE Journal* 48(9):62–69.

Torcellini, P., S. Pless, M. Deru, B. Griffith, N. Long, and R. Judkoff. *Lessons Learned from Case Studies of Six High-Performance Buildings.* 2006. NREL Report No. TP-550-37542. Golden, CO: National Renewable Energy Laboratory. www.nrel.gov/docs/fy06osti/37542.pdf.

USGBC. 2005. LEED NC Indoor Environment Quality Credit 6.1, "Controllability of Systems: Lighting." Washington, DC: U.S. Green Building Council.

USGBC. 2005. LEED NC Sustainable Sites Credit 8, "Light Pollution Reduction." Washington, DC: U.S. Green Building Council.

WEB SITES

3E Plus (Insulation Thickness Computer Program)
    www.pipeinsulation.org
AACA—Air Conditioning Contractors of America
    www.aaca.orgAAMA—American Architectural Manufacturers Association
    www.aamanet.org
Advanced Lighting Guidelines
    www.newbuildings.org/lighting.htm
AIA—American Institute of Architects
    www.aia.org
AIA Committee on the Environment Top Ten Awards
    www.aiatopten.org

ANSI—American National Standards Institute
www.ansi.org
API—Alliance for the Polyurethanes Industry
www.polyurethane.org
ASHRAE—American Society of Heating, Refrigerating and Air-Conditioning Engineers, Inc.
www.ashrae.org
ASTM—American Society for Testing and Materials
www.astm.org
Building Energy Codes Program, EERE, DOE
www.energycodes.gov
CBECS—Commercial Buildings Energy Consumption Survey, EIA
www.eia.doe.gov/emeu/cbecs/contents.html
CRRC—Cool Roof Rating Council
www.coolroofs.org
DesignLights Consortium
www.designlights.org
DOE—U.S. Department of Energy
www.energy.gov
EIA—Energy Information Administration
www.eia.doe.gov
EERE—Energy Efficiency and Renewable Energy, DOE
www.eere.energy.gov
ENERGY STAR®
www.energystar.gov.
EPS Molders Association
www.epsmolders.org
High Performance Buildings Database, EERE Buildings Program, DOE
www.eere.energy.gov/buildings/database/
IESNA—Illuminating Engineering Society of North America
www.iesna.org
Know-How Guide for Retail Lighting
www.designlights.org/downloads/retail_guide.pdf
LBNL—Lawrence Berkeley National Laboratory
www.lbl.gov
LEED—Leadership in Energy and Environmental Design
www.usgbc.org/LEED
Lessons Learned from Case Studies of Six High-Performance Buildings, NREL
www.nrel.gov/docs/fy06osti/37542.pdf.
LRC—Lighting Research Center
www.lightingresearch.org
NAIMA—North American Insulation Manufacturers Association
www.naima.org
NBI—New Buildings Institute
www.newbuildings.org
NEEP—Northeast Energy Efficiency Partnerships
www.neep.org
NEMA—National Electrical Manufacturers Association
www.nema.org

NFRC—National Fenestration Rating Council
www.nfrc.org
NREL—National Renewable Energy Laboratory
www.nrel.gov
PIMA—Polyisocyanurate Insulation Manufacturers Association
www.polyiso.org
RPI—Rensselaer Polytechnic Institute
www.rpi.edu
"Tips for Daylighting with Windows," *Daylight and Windows*
http://windows.lbl.gov/daylighting/designguide/designguide.html
USGBC—U.S. Green Building Council
www.usgbc.org
The Whole Building Design Guide
http://wbdg.org/
XPSA—Extruded Polystyrene Foam Association
www.xpsa.com

For more information or to provide feedback on the *Advanced Energy Design Guide* series, please visit www.ashrae.org/aedg.